Changing Sea Levels

Effects of Tides, Weather and Climate

David Pugh
Southampton Oceanography Centre, UK

CAMBRIDGE UNIVERSITY PRESS

PUBLISHED BY THE PRESS SYNDICATE OF THE UNIVERSITY OF CAMBRIDGE
The Pitt Building, Trumpington Street, Cambridge, United Kingdom

CAMBRIDGE UNIVERSITY PRESS
The Edinburgh Building, Cambridge, CB2 2RU, UK
40 West 20th Street, New York, NY 10011–4211, USA
477 Williamstown Road, Port Melbourne, VIC 3207, Australia
Ruiz de Alarcón 13, 28014 Madrid, Spain
Dock House, The Waterfront, Cape Town 8001, South Africa

http://www.cambridge.org

First published 2004

Printed in the United Kingdom at the University Press, Cambridge

Typefaces Times NewRoman 10/13 pt. and StoneSans *System* LaTeX 2ε [TB]

A catalogue record for this book is available from the British Library

Library of Congress Cataloguing in Publication data
Pugh, D. T.
Changing sea levels: effects of tides, weather and climate / David Pugh.
p. cm.
Includes bibliographical references and index.
ISBN 0 521 82532 6 – ISBN 0 521 53218 3 (paperback)
1. Sea levels. 2. Tides. 3. Storm surges. 4. Climatic changes – Environmental aspects.
I. Title.
GC89.P84 2004
551.45′8 – dc22 2003055752

ISBN 0 521 82532 6 hardback
ISBN 0 521 53218 3 paperback

Contents

The colour plates are situated between pages 82 and 83.

Preface

Our scientific conference in the Maldives on climate and sea level change was going well. As a break we were taken to meet a group of local people to hear their concerns for the future of their beautiful, yet low-lying and vulnerable tropical island Republic. A little to the side of the main demonstration of speeches and banners, a small boy held up high his homemade poster. It declared 'Down with sea level rise'. Worldwide popular concern about possible global warming and sea level rise has been expressed in many ways, but rarely as simply or as effectively.

There is a sea level problem. It may affect, and should concern, us all. Political, economic and social responses need to be guided by scientific evidence and reliable interpretation of the processes involved.

This publication is an introduction to the necessary scientific assessments. It looks at sea level change in terms of the three main causes – astronomical tides, weather and climate trends. It is aimed at undergraduate students of all levels. More advanced students are guided to extend their studies by wider reading. It will also interest and inform professionals in many fields including hydrographers, coastal engineers, geologists, biologists and, perhaps, coastal planners, marine lawyers and economists.

This book began as a development from my earlier, still widely available but now out-of-print, book on *Tides, Surges and Mean Sea-Level*. The older book contains more detail on many topics, but this shorter and more basic book includes much new material, especially on satellite altimetry and climate change effects. My aim has been to reduce the mathematics to the minimum level necessary to describe the processes. More details are given in an appendix, and on the website that accompanies this book (http://publishing.cambridge.org/resources/0521532183/). This website will also update the Further reading sections at the end of each chapter. It includes detailed answers to the questions, which are intended to provoke discussion and extend the text, rather than to ensnare the reader in difficult calculations.

Starting to write a book is much easier than converging on the final product. My progress has been encouraged, advised and helped by many colleagues and friends. Foremost among these are Sylvia Allison, Isabel

Goncalves Araújo, Kate Davis, George Maul, Ana Paula Teles and, as ever, Philip Woodworth.

Others too have helped in many ways, providing additional information and illustrations. These include Carl Amos, Peter Challenor, Lee Harris, John Hunter, Christian Le Provost, Alex Mustard, Adrian New, Jonathan Sharples, Jose da Silva, Robert Smith, Helen Snaith and Alan Suskin. I am particularly grateful to students at Southampton University and the Florida Institute of Technology who helped road test early versions. These include Yasser Abualnaja, Abdullah Al-Subhi, William Carter, Frank Lesley, Natasha Labaume, Jeffrey Simmons and Ivan Haig. Specific acknowledgements are given as appropriate in the text.

Writing a book is a selfish activity. Once again I am indebted to Carole for allowing and even encouraging this indulgence. It has been written at home, in hotels and on planes travelling among five continents. As a result the examples and illustrations used are truly global, as are the issues this book addresses.

Acknowledgements

The following figures are reprinted with permission from Elsevier, Copyright (2001). From *Sea Level Rise: History and Consequences* (edited by Douglas, Kearney and Leatherman), Figures 7.1, 7.7, 7.8 and 7.13 which appear in the original publication on pages 41, 173, 10 and 54. From *Satellite Altimetry and Earth Sciences* (edited by Fu and Cazenave), Figures 3.5, 4.16a and 7.3 which appear in the original publication on pages 271, 292 and 339.

Symbols

a	earth radius
c	wave speed; $(gD)^{\frac{1}{2}}$ in shallow water
C_a	speed of sound in air
C_e	speed of electromagnetic wave
C_D	dimensionless drag coefficient
D	water depth
d_l, d_s	declinations of the moon and sun
f	Coriolis parameter $f = 2\omega_s \sin\phi$
F	a form factor which describes the relative importance of diurnal and semidiurnal tides at a particular location
f_n	nodal amplitude factor for harmonic constituent n
g	gravitational acceleration
G	universal gravitational constant
g_n	phase lag of harmonic constituent n on the local Equilibrium Tide. Relative to the Equilibrium Tide at Greenwich, the symbol used is G_n (usually expressed in degrees)
H_n	amplitude of harmonic constituent n of tidal levels. H_o is the amplitude of a Kelvin wave at the coast
l	length variable
m_e, m_l, m_s	mass of earth, moon, sun
$O(t)$	observed series of sea levels
P	general pressure variable
P_A	atmospheric pressure at the sea surface
$Q(z)$	probability of a level z being exceeded in one year
r	distance, variously defined
R	Rossby radius
$S(t)$	meteorological surge component of sea level
t	time
$T(t)$	tidal component of sea level
T_L	design life of an engineering structure
u	current speed (often used for tidal currents)
u_n	nodal phase factor for harmonic constituent n

V_n	nodal astronomical phase angle of harmonic constituent n in the Equilibrium Tide, relative to the Greenwich Meridian
W	wind speed
$Z_0(t)$	mean sea level
ζ	displacement of water level from the mean
ρ	seawater density
σ	standard deviation of a time series
ϕ	latitude
ω_n	angular speed of constituent n
ω_0 to ω_6	angular speeds of astronomical variables (see Table 3.2)
ω_s	angular speed of the earth's rotation on its axis relative to a fixed celestial point ($\omega_s = \omega_0 + \omega_3 = \omega_1 + \omega_2$)

Chapter 1
Introduction and measurements

Sea levels are always changing, for many reasons. Some changes are rapid while others take place very slowly. The changes can be local or can extend globally. In this introductory chapter we establish some basic ideas of sea level change before looking at the various processes involved in more detail.

The first part of this chapter is an introduction to sea level science as we develop it in this book. It explains the importance of understanding sea level changes and outlines how sea levels are affected by a wide range of physical forces and processes. This is followed by a brief account of the development of ideas on the reasons why sea levels change. The second part is about ways of measuring sea levels. All studies of sea level should be based on reliable measurements over as long a period as possible: we outline the many methods that are available, and discuss their various advantages and disadvantages.

1.1 Background

Living by the sea has many benefits. It offers possibilities of trade and travel, and increasingly of water-based recreation. Natural geological processes have often conspired to create flat and fertile land near to the present sea level, to which people are drawn or driven to settle because the living is usually agreeable.

But there are risks. Sometimes high tides and storms combine to flood low-lying coastal regions causing local damage. Throughout history, humankind has adapted to periodic coastal flooding, but as our cities and our patterns of coastal development become more intricate, populated and interdependent, we become more and more vulnerable to disasters. The rural response of driving cattle to higher ground for the duration of a flood is much easier than the urban complexity of rebuilding complete sewerage and transport systems. In extreme cases the delicate infrastructure of coastal cities may be destroyed, with disastrous long-term consequences.

In November 1966, St Mark's Square in Venice was covered by more than one metre of water. It has been reported that in the first decades of the twentieth century, St Mark's Square was invaded by water seven times per year. By 1990, flooding occurred on average more than forty times a year. With a further 30 cm increase in average sea levels, St Mark's Square would be flooded on average 360 times a year, until defences are built to provide a higher level of protection.

In the long-term, defence is not always possible, nor is it always easy to justify protection in strict economic terms. For example, the Maldives Islands in the Indian Ocean are on average less than 2–3 m above sea level and the Government fears that the Republic's very survival may be threatened by global increases in sea level. Elsewhere, the delta regions of Bangladesh and Egypt are among the most densely populated on earth and the people who live there are especially vulnerable. Protection in these cases will be very difficult and expensive.

Humankind is only one of the biological species that has adapted to the challenging environmental conditions for survival in the coastal zone. Rocky shores are colonised in horizontal bands or zones by plants and animals that have adapted, and that can tolerate different degrees of immersion and exposure. Coral reefs, mangrove swamps and salt marshes are other areas of similar intense coastal biological activity and zonation.

To predict future changes and the impacts of human activity, it is necessary to have a full understanding of all the factors that influence sea levels at the coast. The first step is to make measurements of sea level over a long period, so that there are firm facts on which to base a scientific discussion.

1.2 Changing sea levels

Anyone who had the patience to measure sea levels at the coast for a whole year would find a very regular and rather unexciting pattern of changes. Figure 1.1 shows a year of monthly measurements at Newlyn, a small fishing and recreational port in the southwest corner of Britain. Newlyn sea levels will often be used in this book as examples for our analyses, because Newlyn has a very long record of accurate measurements. Figure 1.2 shows the Newlyn tide gauge location and the

Figure 1.1. A year of sea level observations at Newlyn, southwest Britain, plotted month by month. The dominant semidiurnal tides and the spring–neap changes in range are evident.

benchmark, which defines the zero level for the measurements. The Newlyn benchmark is special: it is used to define the zero level for all British land levelling, based on the average, or mean sea level, over a long period (1915–21). The gauge, which was first installed for this purpose, has been well maintained since to establish a fundamental sea level data series.

Sea levels are changed by factors that extend over a wide range of space and time scales. Figure 1.3 is a space–time map of the main factors. It is drawn in terms of the time scales and the distance scales over which these factors operate. The approximate ranges of the variations associated with each effect are shown; the shapes plotted are only indicative, but note that tidal effects appear as narrow lines at times of one day and half a day. These are the diurnal and semidiurnal tides. Over long geological times, to the right of the diagram, many tectonic processes have changed land and sea levels; in the bottom left-hand corner, over much shorter periods of seconds, there are local wind waves.

In this book we will concentrate mainly on the sea level changes in between these two extremes – those that last from minutes to tens

Figure 1.2. The location (a) and harbour details (b) of the Newlyn tide gauge. Newlyn is separated from the Atlantic Ocean by 200 km of shallower continental shelf. The fundamental benchmark for land levelling datum definition in Britain is located alongside the gauge (c).

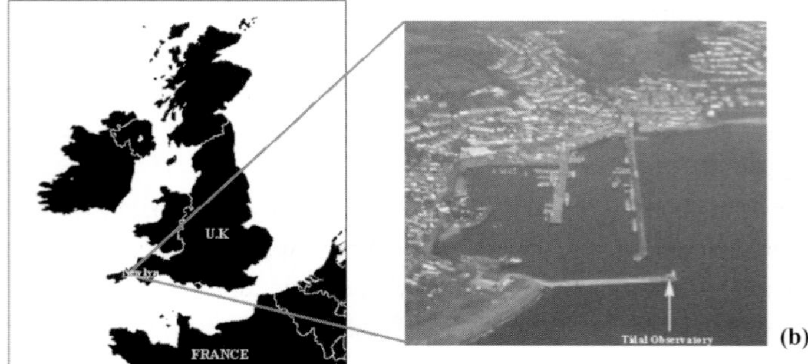

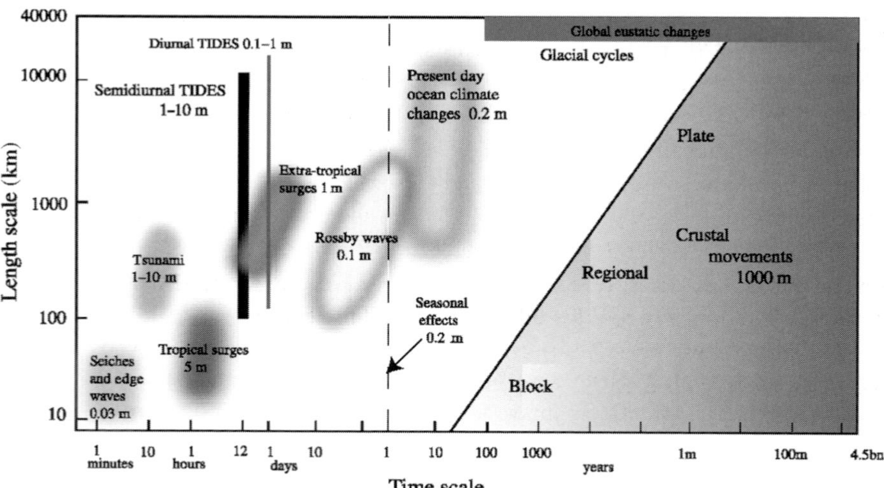

Figure 1.3. A map of the factors that change sea levels in space and time, with typical ranges in metres. Each of the factors discussed in this book occupies a different position on the map. Small-scale rapid changes are in the bottom left-hand corner.

of years. Within this range, sea level records are usually dominated by twice-daily oscillations due to the tides, although there are also seiches, tsunamis, weather effects (surges) and seasonal cycles. The average sea level about which these changes occur is generally called the *mean sea level*. Initially, for some purposes we can consider the mean sea level to be constant, but it does change a little from year to year, and substantially over much longer periods, as we shall show later. We will be looking at all of these processes in more detail, but in the remainder of this section we will describe some of their characteristics as a more general introduction.

For most purposes it is useful to regard the observed sea level as the combined result of three main factors:

$$\text{Observed sea level} = \text{tidal level} + \text{surge level} + \text{mean sea level}$$

These will be considered in turn in this book.

The two main *tidal* features of any sea level record (Figure 1.1) are the *range*, measured as the height between successive high and low levels, and the *period*, the time lapse between one high (or low) level and the next high (or low) level. The tidal responses of the ocean and the responses of the local seas to the forcing of the moon and sun are very complicated, and tidal ranges vary greatly from one site to another.

Nevertheless, in most of the world's oceans the dominant tidal pattern is similar. Each tidal cycle takes an average of almost 12.5 hours, so that two tidal cycles occur for each transit (passage) of the moon through the local longitude. Because each tidal cycle occupies roughly half of a day, this type of tide is called *semidiurnal*. Semidiurnal tides have a range that typically increases and decreases over a fourteen-day period. The maximum ranges, called *spring tides*, occur a day or two after both new and full moons, whereas the minimum ranges called *neap tides*, occur shortly after the times of the first and last lunar quarters. This relationship between tidal ranges and the phase of the moon is due to the additional tide-raising attraction of the sun, which we will discuss in Chapter 2. In Chapter 3 we will develop the idea of adding together several partial tides to represent the observed sea level variations at any particular location, as a tool for tidal analysis and prediction.

In many places, for example at San Francisco on the west coast of the USA, tides with a one-day period, called *diurnal* tides, are similar in magnitude to the local semidiurnal tides. This composite type of tidal regime is called a *mixed* tide. The largest diurnal tides are found in northern Australia and in the Arafura Sea between Australia and New Guinea. Other large diurnal tides are found around Behai Gang in the Gulf of Tongking, China. The dynamics of the ocean response to astronomical tidal forcing, which leads to such a variety of tidal patterns, will be discussed in Chapter 4.

Astronomical forces acting on the major oceans of the world generate and energise the tides. From there the tides spread as waves to the surrounding shallower shelf seas. The tidal ranges on the relatively shallow continental shelves are usually larger than those of the oceans, and it is here that the tides have their biggest impact. Chapter 5 will deal with the behaviour of tides in shallow water and near the coast.

Tidal currents, often called tidal streams, have similar variations. Semidiurnal, diurnal and mixed tidal currents occur, usually having the same characteristics as the local changes in tidal sea levels, but this is not always so. For example, the currents in the Singapore Strait are often diurnal in character while the elevation changes are semidiurnal. The reason for this apparently strange behaviour will be made clearer in Chapter 4. The strongest tidal currents are found in shallow water or through narrow channels that connect two seas, such as the currents through the Straits of Messina between Sicily and the Italian mainland.

The regular and predictable pattern of the tides is slightly (but sometimes spectacularly) altered by the weather, as atmospheric pressure and the winds act on the sea surface. These weather effects (called surges) will be discussed in Chapter 6. Historically, extreme storms have caused many disastrous coastal floods due to the coincidence of large weather-induced surges and large or even moderate high tides. For example, in November 1885, New York was inundated by high sea levels generated by a severe storm that also caused flooding at Boston. More than 6000 people were drowned in September 1900 when the port of Galveston in Texas was overwhelmed by waters that rose more than 4.5 m above the mean high water level, as a result of hurricane winds blowing at more than 50 $\mathrm{m\,s^{-1}}$ for several hours. Even these disasters were surpassed by the Bangladesh tragedy of 12 November 1970 when winds raised sea levels by an estimated 9 m. More recently, in October 1999, 10 000 people were killed in Orissa, India, by a 7–8 m surge.

Tsunamis, generated by submarine earthquakes or landslides, are another cause of rare but sometimes catastrophic flooding, particularly for coasts around the Pacific Ocean. Tsunamis are sometimes popularly called 'tidal waves' but this is misleading, because tidal forces do not generate them, nor do they have the periodic character of tidal movements. The naturalist Charles Darwin in *The Voyage Of The Beagle* describes how, shortly after an earthquake on 20 February 1835, a great wave was seen approaching the Chilean town of Concepcion. When it reached the coast it broke along the shore in 'a fearful line of white breakers', tearing up cottages and trees. The water rose to 7 m above the normal maximum tidal level. The largest earthquake of the twentieth century, off southern Chile in 1960, caused huge tsunamis locally and throughout the Pacific Ocean. Tsunamis often set up local oscillations of semi-enclosed sea and basins, called *seiches*, which are also discussed in Chapter 6; more commonly seiches are triggered by winds or internal ocean tides.

Average or *mean sea levels* have generally increased worldwide by about 0.15 m in the past century, due to melting of grounded ice (ice sheets and glaciers) and to the thermal expansion of warming seas. There are local variations: in polar regions sea levels are falling relative to the land, because the land itself is still rising as it recovers from the loading of glaciers thousands of years ago. The local mean sea level at a site is always defined relative to a fixed benchmark, which is protected if possible from movement, to give long-term stability. There is now much popular and scientific interest in how mean sea level will change in this century, in response to enhanced greenhouse global warming; we will return to discuss this in detail in Chapters 7 and 8.

Knowledge of the probability of occurrence of extreme events, as discussed in Chapter 8, is an essential input to the safe design of coastal defences and other marine structures. If mean sea levels increase, extreme sea levels will occur more often. Dramatic extreme sea levels and the resulting coastal flooding are rare events, but there is always a continuous background of sea level changes due to the weather, which raise or lower the observed levels compared with the predicted tidal levels.

1.3 Historical ideas

The link between the moon and tides has been known since very early times. Sailors had a very practical reason for developing this understanding, particularly for their near-shore navigation in the small ships of those times. A more scientific explanation of the links between tides and the movements of the moon and sun evolved much later. Many eminent scientists have been involved in this scientific development.

Johannes Kepler (1596–1650), while developing laws to describe the orbits of the planets around the sun, suggested that the gravitational pull of the moon on the oceans might be responsible for tides. Isaac Newton (1642–1727) took this idea much further. Almost incidentally to the main insights of his *Principia* published in 1687 – the fundamental laws of motion and the concept of universal gravitational attraction between massive bodies – Newton showed why there are two tides a day and why the relative positions of the moon and sun are important. His contemporary, Edmond Halley (1656–1742), made systematic measurements and prepared a map of tidal streams in the English Channel. Halley had encouraged Newton, paid for the publication of *Principia* himself and prepared an account of the tides based on Newton's work.

Newton's fundamental understanding has been extended and improved by many other scientists, but it remains the basis for all later developments. Daniel Bernoulli (1700–82) published ideas about an Equilibrium Tide which we shall look at in detail in Chapter 2. The Marquis de Laplace (1749–1827) developed theories of a dynamic ocean response to tidal forces on a rotating earth, and expressed them in periodic mathematical terms. Thomas Young (1773–1829), while developing his theory on the wave characteristics of light, showed how the propagation of tidal waves could be represented on charts as a series of co-tidal lines.

The first operational automatic tide gauge and stilling-well system for measuring sea levels was installed at Sheerness in the Thames Estuary in 1831 to provide continuous sea level data. These measurements in turn stimulated a new enthusiasm for tidal analysis and the regular publication by British authorities of annual tidal predictions to assist mariners in planning safer navigation. Even before the official tables, tidal predictions were published, sometimes based on undisclosed formulae, for example those of the Holden family in northwest England.

Lord Kelvin (1824–1907) showed in detail how tides could be represented as the sum of periodic mathematical terms and designed a machine that applied this idea for tidal predictions. He also developed mathematical equations for the propagation of tidal waves on a rotating earth, in a form known as *Kelvin waves*. In 1867 the Coastal Survey of the United States took responsibility for the annual production of official national tide tables for the USA. Soon most major maritime countries around the world began to prepare and publish regular annual official tide tables.

Meanwhile, other factors that influence sea level changes were being investigated. James Clark Ross (1800–62) confirmed the already-known link between higher atmospheric pressures and lower sea levels, known as the inverted barometer effect, by sea level measurements when trapped in the ice during the Arctic winter of 1848–49. Earlier, Ross had helped establish a tide gauge benchmark in Tasmania as a datum for scientific mean sea level studies during his voyage of exploration in the Southern Ocean. Establishing these fundamental fixed datum levels was done on the advice of the German geophysicist Alexander Von Humboldt (1769–1859).

Throughout the twentieth century a series of scientific and technical advances has brought us to the current state of being able to map and model ocean and shelf tides in great detail, using satellite altimeters and the processing power of modern computers. Today one of the highest priorities is to understand and reliably anticipate changes in mean sea level and flood risks, particularly those that may be due to global climate change.

Our scientific understanding and our ability to predict future changes depend on the collection of high-quality sea level measurements, made in a variety of ways by different kinds of instruments. In the next section we will consider some of the basic principles on which these instruments operate and some of their individual advantages and disadvantages.

1.4 Measuring sea levels

When measuring sea levels, the aim is to measure the vertical distance between the average surface of the sea and a fixed datum level. We need to smooth out the transient effects of wind waves, which have periods of only a few seconds, to get the averaged sea levels. Measuring sea level is much more difficult than measuring in a laboratory because of difficulties due to waves, corrosion, biofouling, site access, site security and long-term reliability.

Table 1.1 summarises the types of instruments commonly used. These vary from the cheap but inaccurate tide pole, through to dedicated satellite systems. Manuals prepared by the Intergovernmental Oceanographic Commission (1985) of UNESCO give fuller details of how to choose and operate a sea level measuring systems.

When choosing a system it is important to consider the purpose of the measurements. Choices about cost, accuracy, location and duration of measurements follow once this basic purpose is established. Shipping operations may be well served by an accuracy of 0.1 m. For scientific sea level studies, an accuracy of around 0.01 m in individual readings, after wave averaging, is generally possible and acceptable. If the small errors are also random, then averaging over several readings leads to the higher resolution and accuracy that are needed for studying long-term mean sea level trends. For these studies, it is essential to check the gauge zero datum level regularly to ensure that it remains stable. In many cases the cost of good measurements is not much greater than the cost of less accurate systems, once the main structures are paid for. Generally it is best to install the best affordable system, as the data can then serve many different purposes.

1.4.1 Datums

We must emphasise the importance of defining and maintaining a clearly defined and stable zero level or datum for sea level measurements. The datum chosen depends on the application. Those most commonly used are summarised in Table 1.2. The chosen datum may vary from the simple tangible local benchmark at a tide gauge through to the much

Table 1.1. *Summary of the characteristics of commonly used methods for measuring sea level.*

Category	Type	Wave averaging	Accuracy	Advantages	Disadvantages
Surface following	Tide pole Float	By eye Stilling well	0.02–0.10 m 0.01–0.05 m	Inexpensive; easy to make and move; robust	Tedious; needs vertical structure; high maintenance
Fixed sensors	Acoustic reflection Radar reflection Pressure	Multiple samples Hydrodynamics and multiple samples	0.005–0.01 m 0.01 m	Robust; low maintenance; low cost; no vertical structure needed	Needs vertical structure; density and wave corrections, high maintenance
Remote and mobile	Satellite	Empirical adjustments	0.01 m	Systematic global coverage: high data rate	Expensive: specialist use only; multiple corrections; misses local storms

Table 1.2. *Datums used for sea-level measurements.*

Datum	How fixed	Applications	Advantages	Disadvantages
Tide gauge benchmark	Local levelling	Single tide gauge reference	Long-term stability; an interconnected group covers for accidental damage	Often damaged; includes vertical land movement
Chart datum	Tidal analysis and levelling	Navigation charts	Level below which sea seldom falls	Varies with tidal range; unsuitable for numerical modelling
Land survey datum (MSL)	Regional levelling and sea level averaging	Approximate geoid for national mapping	Horizontal surface transferred by levelling	Systematic errors in conventional levelling
Geocentric coordinates	Analysis of satellite orbits	Altimetry, GPS	A geometric framework; detects local movements, e.g. of TGBMs	Small changes as mass, e.g. ice, redistributes
Geoid	Satellite orbits and modelling	Ocean circulation	True horizontal surface; absolute level for ocean dynamics	Needs special satellite gravity mission

more abstract ellipsoid of revolution used for referencing satellite orbits (see Figure 1.9a) and the geoid.

The *tide gauge benchmark* (TGBM) (see Figure 1.2) is a stable surface or mark near a gauge, to which the gauge zero is referred. It is connected to a network of local auxiliary benchmarks to check local stability and to guard against accidental damage. From time to time these local marks are inter-connected using standard surveying methods to check the stability of the TGBM and the local network. Although the benchmark itself is usually well above sea level to allow easy access, the levels may be referred to a defined plane that can be several metres below the benchmark. This has the advantage that all measured sea levels are positive above the datum and values increase as sea level rises. Very rarely, authorities measure sea levels downwards from a tide gauge datum so that the values decrease as sea level rises.

A *chart datum* is used for nautical charts and tidal predictions. Normally the datum approximates the lowest astronomical tide, the lowest level that can be predicted to occur under average meteorological conditions and under any combination of astronomical conditions. For seafarers this has the important safety factor that the water navigation depths shown on the charts are always the least possible. Predicted tides can always be added to these depths. This definition in terms of the tidal range means that the chart datum is not a horizontal surface. It moves up and down as the tidal range changes along the coast, but it may be considered horizontal over a limited local area.

A *land survey datum* is often defined from calculations of mean sea level over a specified period. For example, for the United Kingdom Ordnance Datum Newlyn (ODN) is the mean sea level at Newlyn measured between 1915 and 1921. Present mean sea levels at Newlyn have risen to about 0.2 m above ODN. In theory the survey datum parallels the geoid (see below), a true horizontal surface, but because of small systematic errors in land levelling techniques, the transfer of the datum from one tide gauge to another can have errors of several centimetres or more.

The *reference ellipsoid* reflects the fact that the earth is not a perfect sphere. It would be nearly spherical if gravitational forces only were involved, but due to its rotation the earth bulges out slightly at the equator and is slightly flattened at the poles. The radius at the equator (6378 km) is 21 km greater than the radius at the poles (6357 km). An ellipsoid rotated about the polar axis is a good fit to this shape. There are different ellipsoid definitions in use. The ellipsoid used by the TOPEX/Poseidon scientists has an equatorial radius of 6378.136 km and a flattening coefficient of 1/298.257. This is used when calculating and correcting the TOPEX/Poseidon orbits as shown in Figure 1.9a. It is also used as a reference level for the geoid.

The *geoid* datum is a true but irregular level horizontal surface. It is the theoretical mean sea level surface, which varies from the ellipsoid of revolution by as much as 100 m due to the uneven distribution of mass within the earth. The geoid, which is favoured by oceanographers who calculate ocean currents, will be discussed in the next chapter and in Chapter 7.

1.4.2 Direct surface measurements

Direct measurements by following the moving sea surface are the easiest to make, and there are many kinds of instrument available.

Tide poles

Tide poles or staffs are cheap, easy to install and may be used almost anywhere. Many harbours have levels engraved in their walls (Figure 1.4), often related to the depth of the harbour entrance or of a dock sill, to show a ship if there is enough water for safe entry. Taking the average level between the crest and the trough levels over a short period, perhaps 20 seconds, averages out waves. The accuracy of the readings can be substantially increased by fitting a transparent hollow tube alongside the pole, connected to the sea at the bottom through a narrower tube, which damps out external waves very effectively. The level in the tube is easily read against the graduated pole, particularly if some dye is mixed into the tube water. A series or flight of poles levelled to the same datum can be used for measuring up a gradually sloping beach. Poles are often the best choice for short-term surveys of only limited accuracy, but the tedium involved and the errors that result if readings are needed over several months makes them unsuitable for long-term measurements.

Float gauges

Over the past 150 years, and until quite recently, float gauges (Figure 1.5) were the standard method of measuring and automatically recording sea levels. The system consists of a float, connected by a system of wire, pulleys and gearing, which moved a pen over a paper chart. Float gauges are still commonly used, but digital recording is now done by connecting the wire and pulley arrangements to electronic counters; data may be transmitted by telephone lines or satellites.

Float gauges require a vertical structure so that the recording system can be mounted above a stilling well. The stilling well is a vertical tube, long enough to cover the whole range of tides at the site. The bottom of the well is closed off except for a small entry for the water to flow in and out and the top is left open. The bottom connection may be either

Figure 1.4. An early example of a tide staff, showing levels above Old Dock Sill datum, at the entrance to Canning Half-Tide Dock, Liverpool (provided by Proudman Oceanographic Laboratory). Modern tide staffs are designed for easy reading.

a small orifice or a pipe inlet. This narrow connection damps out rapid disturbances such as waves and ship wash (gauges are often located in busy harbours), but adjusts to longer period variations such as tides. Except at very exposed sites an orifice-to-well diameter of 0.1 (an area ratio of 0.01) gives satisfactory results. The well diameter is typically 0.3–0.5 m.

Copper is often used to prevent marine growth blocking the narrow connections to the sea. Where icing is a winter problem, a layer of kerosene may be poured over the seawater in the well. The float should be as large as possible to give maximum force to overcome friction in

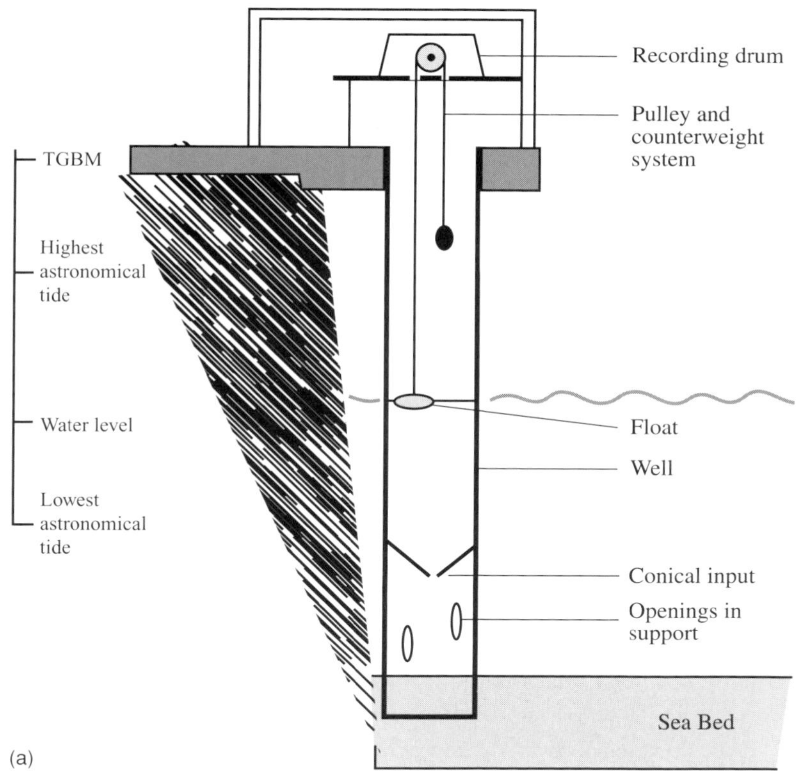

Figure 1.5. (a) The standard stilling-well system used by the USA National Ocean Service. The orifice diameter is typically 10 per cent of the well diameter, except at very exposed sites. The well diameter is typically 0.3–0.5 m. (b) The tide gauge at the entrance to the Aveiro Lagoon, Portugal; the inset shows the traditional chart gauge inside the building, located on top of the stilling well (photograph by Isabel Goncalves Araújo).

the recording system, but must move freely and not touch the side of the well. Routine daily and weekly checks are essential. The recorder zero must be checked periodically against an external tide pole. Some gauges have a datum switch fitted, either inside or outside the stilling well. This switch is triggered when the water level reaches a fixed height above the datum (often near the mean sea level) so that corrections can then be made for any slippage in the float and wire system.

Apart from occasional mechanical difficulties, the accuracy of stilling-well systems is limited by two factors: the systematic water density differences (salinity and temperature) between the inside and outside of the well leads to differences in the levels; also, where there are strong currents, the water flowing past the submerged well orifice can lead to a drawdown in the well levels.

Other types of surface-following sensors are possible. Offshore, where there are no fixed structures, satellite positioning systems, notably the Global Positioning System (GPS), have been placed on floating, loosely moored buoys to measure sea levels. Accuracies of a few centimetres are possible by averaging over several readings to eliminate waves. The absolute accuracy will depend on the way the buoy responds to the waves, and over the longer term, the effects of biofouling and leakage on the level at which the buoy floats in the water.

1.4.3 Fixed sensors

This category includes sensors that measure their distance from the moving water surface by reflected signals and sub-surface sensors, which measure the changing water head pressure.

Acoustic tide gauges

The time taken by a pulse of sound to travel from the source to a reflecting surface and back again is a measure of the distance from the source to the reflector. The travel time is given by:

$$t = \frac{2l}{C_{\mathrm{a}}} \tag{1.1}$$

where l is the distance to be measured and C_{a} is the velocity of sound in air. Corrections must be made for variations in C_{a} with air temperature, pressure and humidity. Acoustic gauges are now commonly used, either replacing the float mechanism in existing stilling wells, or as separate installations.

The US National Oceanic and Atmospheric Administration (NOAA) and other authorities began installing acoustic gauges with sound-transmitting tubes in the early 1990s (Figure 1.6). The acoustic gauge

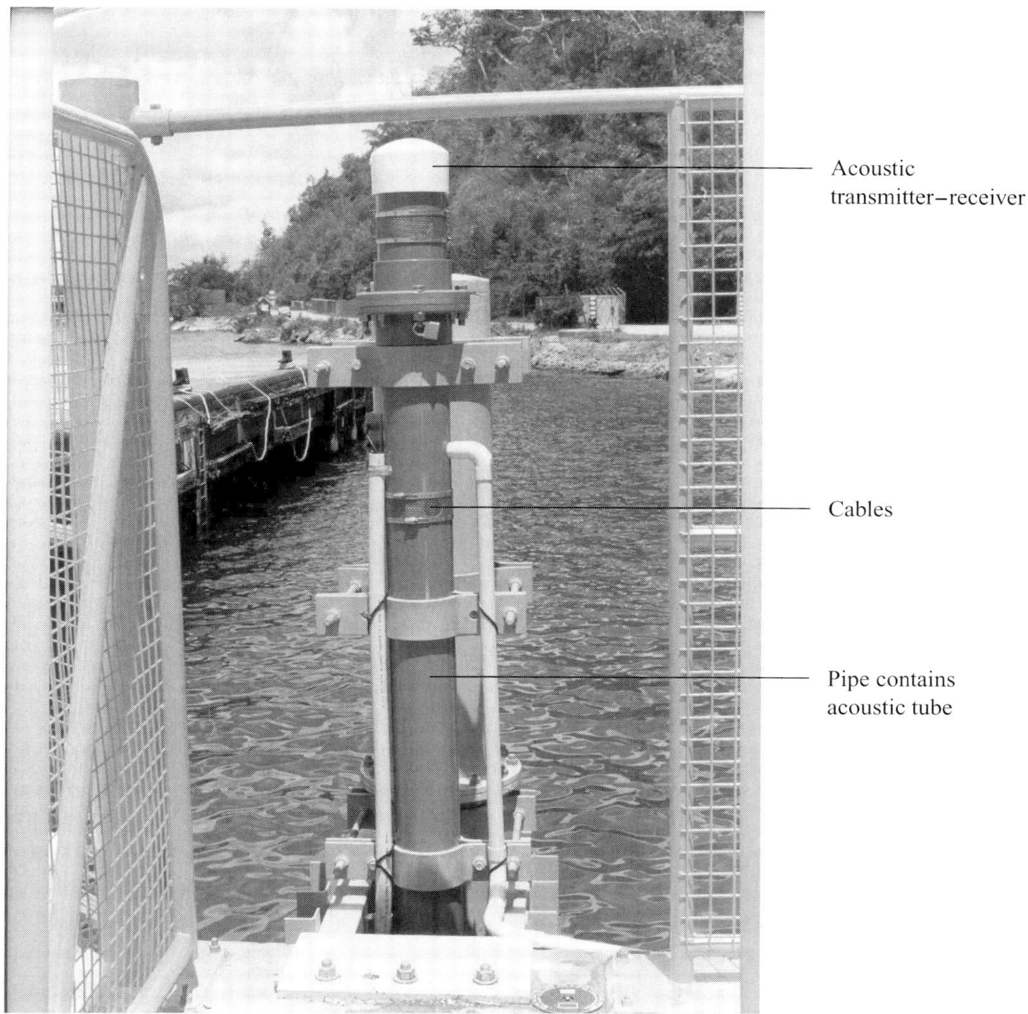

Figure 1.6. The acoustic Aquatrak tide gauge at Port Vila, Vanuatu. The acoustic pulse transmission head sits over the guide tube. Waves are removed by averaging one-second readings over a three-minute period (provided by the Australian National Tidal Facility).

sends a pulse down a channel, a 1.3 cm diameter sounding tube, towards the water surface and measures the travel time of the reflected signal. The sounding tube has an acoustic discontinuity, sometimes a small hole, at a known distance from the top of the tube which gives a fixed second reflection point. The elapsed times for the return of the reflected pulses from the water surface and from the fixed point are recorded. The travel time to the fixed reflection point is used as a calibration, automatically allowing for changes in the speed of sound due to air temperature and pressure effects.

The acoustic sounding tube is mounted inside a 0.15 m diameter protective well which has a symmetrical 0.05 m diameter double cone orifice at the bottom. The protective well is more exposed to local dynamics than the traditional stilling well used for float gauges and only partly filters out wind waves. Rapid sampling and averaging is used to fully remove the waves. In areas of high-velocity currents and high-energy sea swell and waves, parallel plates are mounted below the orifice to reduce the drawdown effects.

Acoustic systems with data transmitted to a central location are widely used in the USA, Australia and in many other countries. In addition to sea level, other parameters such as water temperature, air pressure, air temperature and wind speed can be logged and transmitted at the same time. In a typical gauge the water levels are measured once per second. Not all data are stored and transmitted. The primary water level measurements are averaged typically over a three-minute period and are stored in the memory every six minutes. To detect a change in water level of 0.01 m the timing of the acoustic pulse travel must be accurate to about 60 microseconds.

Some acoustic gauges have been mounted directly above the open sea surface without any mechanism for damping waves. This method depends on very rapid sampling for wave averaging. There is some advantage of not having to provide a protective stilling-well system, but this is offset by the problems of getting a measurable return signal from a sea surface that is sloping in different directions as waves pass.

Acoustic gauges can be very sensitive to air temperature changes, if corrections are not made. An alternative is to transmit a pulse of radar (as do altimeters) rather than sound, as radar transmission times are not sensitive to air temperature changes.

Pressure measuring systems

The pressure at some fixed point below the sea surface is related to the overlying water level:

$$P = P_{\mathrm{A}} + \rho g D \tag{1.2}$$

where P is the measured pressure at the instrument depth, P_{A} is the atmospheric pressure acting on the water surface, ρ is the mean density of the overlying column of seawater, g is the gravitational acceleration and D is the depth of the water column above the transducer.

A gas bubbling system such as that shown in Figure 1.7 is a simple tide gauge with good overall accuracy and datum stability. Compressed air or nitrogen gas from a cylinder is reduced in pressure through one or

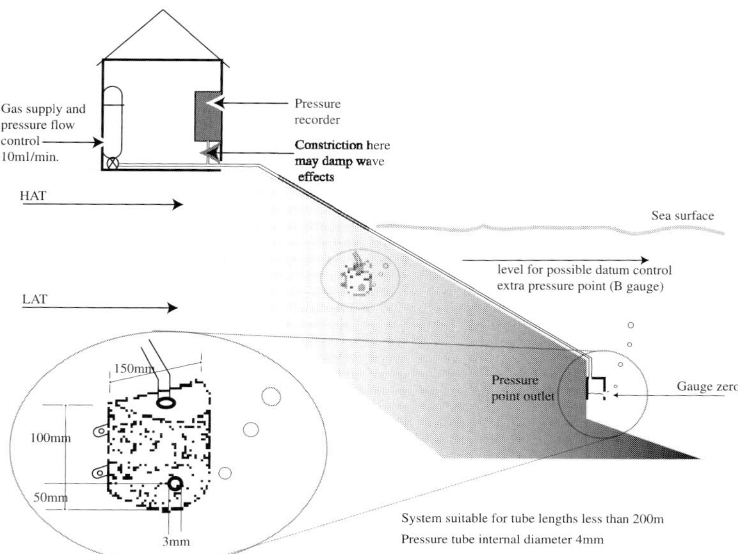

Figure 1.7. A basic pneumatic bubbler system for tube lengths less than 200 m. The pressure at which bubbles escape from the underwater pressure point gives the water head pressure. This is measured on the pressure recorder ashore. Note the possible location of a second, more accessible pressure point for regular datum checking.

two valves so that there is a small steady flow down a connecting tube to escape through an orifice in an underwater canister, called a pressure point. The level of the orifice is the gauge zero. At this underwater outlet, for low rates of gas escape, the gas pressure at the pressure point is equal to the water pressure P. This is also the pressure transmitted up the tube for measuring and recording.

The normal procedure is to measure the pressure using a differential transducer, which responds to the difference between the system pressure P and the atmospheric pressure P_A, so that only the difference $(P - P_A)$, the water head pressure, is recorded. If g and ρ are known, the water level relative to the pressure point orifice datum may be calculated from Equation (1.2). For most sites a suitable constant value of water density ρ may be fixed by observation. However, in estuaries the density may change significantly during a tidal cycle, in which case an increase of density with water level can be included in the processing and calculations.

The underwater pressure point is a critical and sometimes neglected part of the system. It is designed to prevent waves forcing water into the connecting tube. If this happened there would be large errors. To avoid errors due to waves, the critical parameter is the ratio between the total volume of air in the pressure point and connecting tube, and the area of the pressure point cross-section: this ratio should not exceed 0.2 m. In practice the connecting tube between the underwater outlet and the recorder achieves some wave damping. For short tube lengths, if waves are a problem, a narrow wave damper can be inserted at the

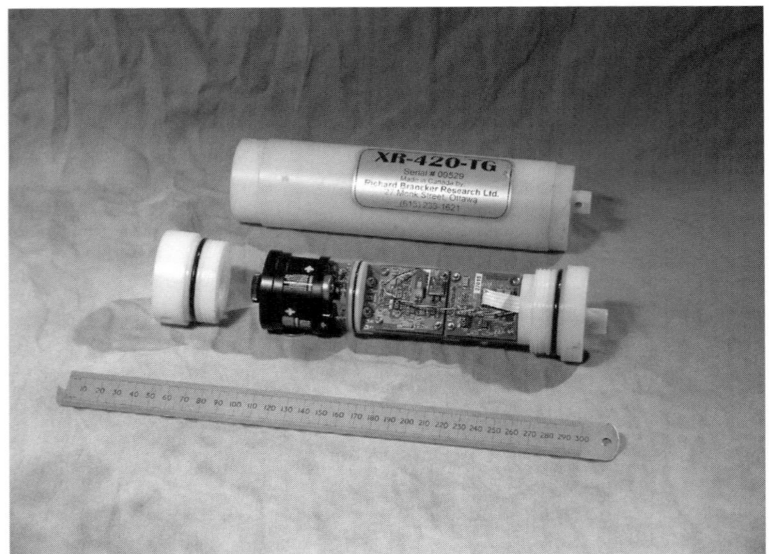

Figure 1.8. Several bottom-mounted sea level pressure gauges are available commercially. The one shown here was developed by Richard Branker Research, for deployment to 100 m depth. These gauges are suitable for measuring seiches and tides, but their zero datum is not stable enough for measuring long-term mean sea level changes. The rule at the bottom of the photograph is 0.3 m long.

entrance to the pressure recording system (see Figure 1.7). For tube lengths up to 200 m the system shown in Figure 1.7 will be accurate to within 0.01 m.

The datum of a bubbling system is the small bleed hole in the side of the pressure point. This must be rigidly fixed to some structure or to the sea bed. Datum control accuracy can be improved by having a parallel system (called a B Gauge) with a second and more accessible pressure point fixed near mean sea level. This serves the same purpose as the datum switch in stilling-well systems, but in this case comparing the differences between the two bubbling gauges when both are submerged checks the datum.

One great advantage of bubbling systems is that they do not need a vertical structure. Separations between the pressure point and the recorder of up to 200 m are relatively straightforward, and with careful design and corrections, connecting tube lengths of 400 m or more are possible. Other advantages of a bubbling system include the stability of a clearly defined datum, and the cheap expendable nature of the vulnerable underwater parts. Even if the connecting tube is accidentally cut, it can be repaired and the system re-established by purging water from the system with a high-pressure air flow for a short period.

Electronic pressure sensors mounted underwater are also used for pressure-measuring sea level systems. Many have been developed for very accurate scientific work at ocean depths. Less expensive gauges are available commercially. Figure 1.8 shows a robust and relatively inexpensive pressure recorder, which can be operated on the sea bed to depths of 100 m. The instrument is programmed directly by computer to sample at selected intervals from seconds to hours. Waves are averaged digitally by measuring over several seconds before recording a value. These instruments are easy to deploy and ideal for short-term tidal studies. However, the sensors are not sufficiently stable to be used for long-term sea level studies; nor are the readings immediately available for operational purposes such as ship navigation.

Some gauges with submerged electronic sensors have their data transmitted ashore through a cable for operational purposes. The disadvantages of these electronic underwater systems are the difficulty of precise datum definition (the zero of electronic sensors tends to drift over time) and the cost of replacing the cable and other relatively expensive underwater components if they become damaged.

The gauges developed by scientists studying tides in the open ocean have specialised bottom-mounted self-contained pressure measuring and recording systems, which have worked for several years in depths of thousands of metres. Deployment and recovery can be controlled by acoustic commands from a surface ship. These instruments cannot be used for mean sea level studies because there is no geodetic datum control. However, the regular tidal movement can be extracted by analysis even from depths of thousands of metres. All pressure sensors are to some extent affected by temperature changes and so the best performance of these specialised gauges is only obtained by careful calibration and correction.

1.4.4 Satellite altimetry

Sea levels measured from satellites are often called *sea surface heights*. These measurements have in two senses revolutionised not only the measurement of sea level, but also methods of analysis and scientific interpretation. The most important satellites for sea level studies are TOPEX/Poseidon, and follow-up missions in the JASON series. These satellites, which orbit the earth every 112 minutes at an altitude of 1336 km and a speed over the ground of 5.8 km s^{-1}, measure the distance to the sea surface by radar every second. The target accuracy is 0.02–0.04 m for a single measurement, but better accuracy can be achieved by averaging over several measurements.

The basic principle of timing a reflected pulse is the same as for the fixed acoustic or radar tide gauge. However, the engineering skills needed to build satellites to measure sea levels to a few centimetres or better are much more complex. Altimetry satellites are radars that transmit

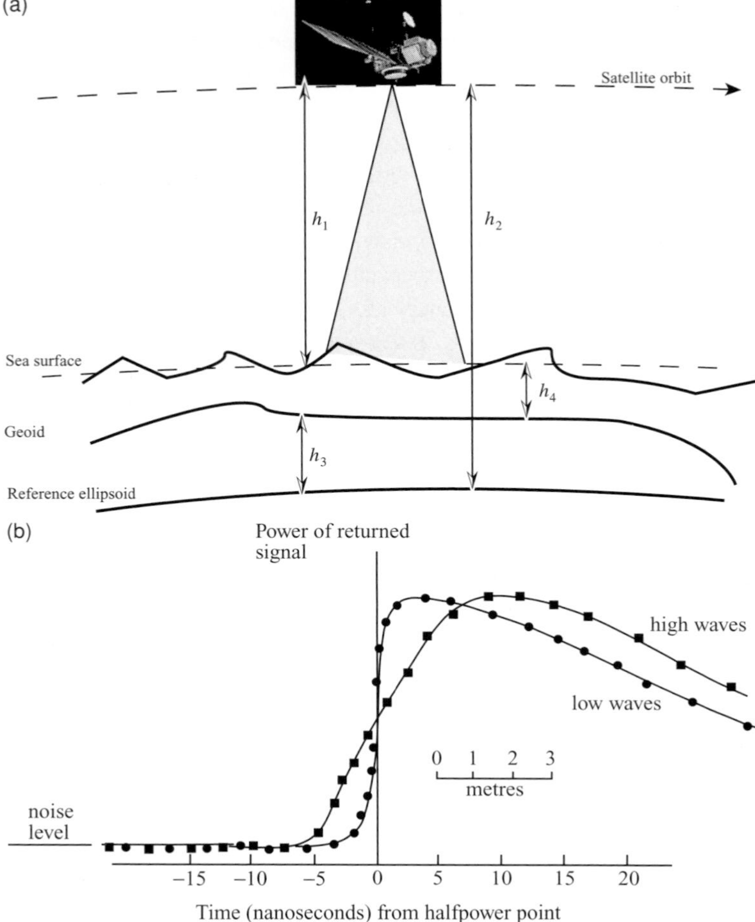

Figure 1.9. (a) Schematic diagram of the parameters required to determine sea level relative to the geoid by satellite altimetry. (b) Typical altimeter return pulses from the sea surface; the return pulse becomes less sharp if the electromagnetic signal is reflected from a sea where there are high waves. The additional horizontal axis shows the sea level equivalent in metres.

very short pulses. The time for the pulse to travel from the altimeter to the sea surface and back is a measure of the height, as in Equation (1.1), but with C_a replaced by C_e, the speed of electromagnetic waves.

Figure 1.9a shows this measured satellite distance (h_1) above the sea surface. This needs to be connected to a fixed datum. For oceanographic work the sea level required (h_4) is measured relative to the geoid. As we shall discuss later (Section 2.5) the geoid is the 'level' surface of equal gravitational potential, that is the level the sea surface would assume if there were no disturbing forces. Satellite elevations (h_2) can be fixed relative to a reference ellipsoid, a simple geometric shape that approximates the shape of the earth. The remarkable techniques for doing this to accuracies approaching 0.01 m are beyond the scope of this account (see the end of this chapter for recommended further reading). The separation between the geoid and the reference ellipsoid (h_3) varies by more than 100 m over the surface of the globe because of uneven mass distribution within the earth. The value of h_4 is calculated as ($h_2 - h_1 - h_3$). Developing precise maps of this separation to centimetre accuracy is a major current research topic.

To achieve a 0.01 m accuracy of the measured satellite distance (h_1) requires the time of travel of the electromagnetic pulse to be accurate within 30 ps (3×10^{-11} s), several orders of magnitude more demanding than for the acoustic gauge. The measurements may also be expressed in terms of the slopes on the sea level surface, with target accuracies of 1 microradian (1 mm in 1 km), or better averaged over longer arcs. In practice a pulse of about 3 ns (3×10^{-9} s) is transmitted and the return signal is fitted to a model curve. Figure 1.9b shows the strength of the reflected electromagnetic wave from the sea surface, over a period of about 40 ns. It also shows the equivalent sea level differences for different pulse return intervals (9 ns is approximately 3 m). The timing of the mid-point of the leading slope of the return signal, shown as zero time in Figure 1.9b, is computed for measuring the sea level. Other information can be extracted from the return signal. The shape of the curve varies with wave height at the reflecting sea surface: for low waves the reflected first return pulse is sharp, but for high waves the return pulse is more diffuse because of reflections from many surfaces. Also, the amplitude of the reflected signal can be related to the wind speed at the sea surface.

To get the best estimate of sea level it is important to understand the various corrections that are necessary for wave height, and for electromagnetic effects in the ionosphere and for gases in the atmosphere. Firstly, the return signal to the altimeter comes only from those sea surface elements perpendicular to the target line. Because of the natural shape of wind waves these elements are more common in the wave trough than in the wave crest, so in waves the average surface from which reflections are received is lower than the average sea level. If waves are high the difference increases; this is known as sea-state bias, and can be up to a few per cent of the wave amplitude. When computing the geoid (Section 2.5) wave bias correction is often one of the biggest causes of uncertainty.

Secondly, the number of free electrons affects the speed of electromagnetic waves in the ionosphere, which varies between day and night, and between winter and summer. The pulse speed depends on the electromagnetic frequency, and so the effect can be corrected for by measuring the different transmission times for pulses at different frequencies. Corrections are also necessary for the effects on C_e of water vapour and dry gases in the atmosphere.

Typical onboard satellite averaging times are 50 ms, and these values are usually further averaged for oceanographic applications over time periods of about a second. These averaged measurements, once per second along the track, give a sample spacing of about 6 km at the sea surface. The exact footprint size, the area of sea surface returning a signal to the altimeter, depends on the significant wave height, and is typically 3–5 km in diameter. This footprint diameter can increase to more than 10 km for very high waves.

The choice of orbit for the satellite is critical. As the earth rotates under the orbit of the satellite, so each successive altimeter track covers a part of the earth's surface slightly further west than the previous one. The height of the orbit controls the period of the satellite rotation about the centre of the earth, and this in turn determines the rate of westward migration of the tracks. This westward movement can be adjusted to give exact repeat orbits after an interval of several days, for a systematic coverage of the oceans (see Figure 3.4b). If the orbit in which the satellite travels is too high, the area measured by the footprint will be too big for scientific work; however, a low orbit height means that the drag on the satellite is increased, which leads to difficulties in precise orbit determination. The orbits for satellite altimeters are set to be nearly circular, at a fixed angle to the plane of the equator. This angle determines the maximum latitude north and south of the equator that the satellite covers; for the fullest coverage, including polar regions, a large inclination to the equator is needed.

The first radar altimeter, Seasat, flew in 1978, followed by Geosat (1985–89). The European Space Agency has flown multipurpose satellites, which included an altimeter, in orbits reaching 81° north and south of the equator (ERS-1, ERS-2 and Envisat).

The joint USA/France TOPEX/Poseidon mission and its successors (1992 onwards) are the first satellites dedicated solely to altimetry. The TOPEX/Poseidon altimeter is a dual-frequency radar instrument. The primary channel is the Ku-band (13.6 GHz), and the secondary channel is C-band (5.3 GHz). The secondary channel allows correction for propagation delays in the ionosphere. The orbit is a 9.9 day exact repeat at an inclination of 66°, 1336 km above the earth. The orbit period is 112 minutes, which means that the ground tracks are about 315 km apart at the equator. With careful averaging, changes of sea level can be measured to 2 cm or better. It is very important that the repeat tracks are identical to within1 km to avoid cross-track slopes in the geoid appearing as apparent changes of sea level with time (see Section 2.5).

Figure 1.10. Jason-1, the follow-up altimeter satellite to TOPEX/Poseidon. Further altimeter satellites in the same series will give valuable data for studying changes in ocean circulation and mean sea level (courtesy of NASA/JPL/Caltech).

The successor altimetry satellite to TOPEX/Poseidon is Jason-1, named after the mythological Greek hero. The specification "1" indicates that Jason-1 should be the first in a series of future altimeter missions, extending the TOPEX/Poseidon data over decades. Jason-1, launched in 2001, is shown in Figure 1.10. Jason-1 also flies in a circular orbit 1336 km above the earth, and has the same orbit and repeat characteristics as TOPEX/Poseidon. Its orbit is similarly inclined at 66° to the equator, so that the polar regions above 66° latitude are not covered; it also has a 9.9 day repeat cycle along the same tracks. The target accuracy is 0.025 m. The 500 kg satellite carries a number of tracking systems including a laser retroflector, GPS and the DORIS system.

Altimetry makes it possible to determine tides, the seasonal cycle of sea level and other changes of the ocean to be determined systematically, globally and with high accuracy. The uses of altimeters to study mean sea level changes will be discussed in Chapter 7. Altimetry, initially a tool for scientific research, is becoming an operational tool for real-time forecasting of ocean circulation. Altimetry data are now routinely assimilated into operational ocean models, and so increasing emphasis is placed on the rapid processing and immediate availability of sea level data. Future missions carrying altimeters that measure sea surface gradients across a 100 km wide swath along the orbit track have been proposed. These would vastly improve the area covered and allow determination of currents, related to the gradients, even in confined coastal regions.

1.4.5 Data reduction and assimilation

There is an increasing demand for immediate access to measured and processed sea level data. Satellite data are now processed and assimilated in ocean models within a few hours, for operational forecasting. The information produced by the altimeter must be immediately adjusted for measured satellite orbit heights and all the other factors discussed above. Similarly, harbour gauges, although much less complicated, transmit data in real-time to assist navigation and shipping movements.

Making measurements of sea level is only part of the data gathering process. Proper long-term management of the data is essential, but is often given low priority. Final checking and corrections to sea level data for mean sea level studies and other scientific interpretation can be done more thoroughly later. Extra time at this stage allows for more careful correction and, if necessary, interpolation of gaps in the data. In all cases, when the data have been processed, it is highly desirable that they are preserved with proper documentation in a permanent digital databank. Data should be easily available, for example for studies of regional and global long-term changes. International agreements exist for the exchange of data between national networks under GLOSS, the Global Sea Level Observing System of the IOC. Further details of data systems are given in the following box.

Sea level data sources

Governments have set up a Global Sea Level Observing System (GLOSS), to which more than 80 countries contribute measurements from their sea level gauges. GLOSS has a core of some 300 gauges which provide an approximately evenly distributed sampling of global coastal sea levels. Another component is the set of gauges for identifying long-term trends in global sea levels. Wherever possible these gauges are linked to benchmarks maintained in geocentric coordinates (Figure 1.11), for example by the Global Positioning System. These geocentric benchmarks allow us to distinguish between local vertical land movement, and sea level changes. The GLOSS altimeter calibration set consists mainly of island stations, away from the effects of shallow water, which are used for calibration of altimetry satellites. The University of Hawaii Sea Level Center collects data from a subset of the GLOSS network in real-time so they are available almost immediately for a range of operational purposes. The Permanent Service for Mean Sea Level, on Merseyside in the United Kingdom has collected and analysed monthly and annual data since 1933 for the science of sea level changes. It holds about 50 000 station-years of data from over 1800 tide gauges worldwide, and from almost 200 national authorities. The data can be accessed through the Internet (www.pol.ac.uk/psmsl).

Figure 1.11. The Global Positioning System (GPS) fixed bench mark maintained by the University of Hawaii at Stanley, Falklands/Malvinas Islands, for measuring long-term vertical land movement. These marks must be fixed to solid rock foundations, and should be connected regularly to local tide gauges.

Tidal beliefs and myths

Ancient people and cultures certainly linked their observations of the tides to the movements and phases of the moon, but there can have been no basis for a scientific explanation of the relationships. Instead, a range of ideas and superstitions developed as 'explanations'. Those who worshipped the moon and sun saw the tides as a terrestrial manifestation of their gods.

Chinese ideas supposed water to be the blood of the earth, with tides the beatings of the earth's pulse; alternatively some thought that the earth's breathing caused tides. One poetic explanation invoked an angel who was set over the seas. When he put his foot in the sea the flow of the tide began; when he raised his foot the ebb tide followed.

The Roman writer, Pliny the Elder, describes how the moon's influence is most strongly felt by those animals that are without blood. Aristotle is credited with the law that no animal dies except when the tide is ebbing. This legend persisted in popular culture, and is referred to in Charles Dickens' *David Copperfield* for the east coast of England. Even as recently as 1595, Parish Registers in the Hartlepool area of the north of England recorded the phase of the tide along with the date and time of each death.

The popular legend about King Canute of England, Denmark and Norway (995–1035) is also found in other traditions. Supposedly, Canute commanded the rising tide to stop – with no effect – and thus convinced his courtiers that he was not all-powerful. Both

Southampton and Bosham on the English south coast have been suggested as the site of this event. A similar legend involves the Welsh Prince, Maelgwyn, in the Dovey Estuary, on the shore of the Irish Sea.

Further reading

Many oceanographic textbooks have one or two chapters summarising tides and sea level changes. The more recent of these include Knauss (1997), Pond and Pickard (1995) and the Open University texts (2000). More detailed discussions of all the topics in this book are given in Pugh (1996). Cartwright (1999) gives an excellent analysis of the scientific history of tides; a more general discussion of sea level science and its place in the overall development of marine science is given in Deacon (1997). Venice sea levels are reported in Zanda (1991).

Useful practical guidance for the choosing and operating of sea level gauges is given in the GLOSS Manuals, published by the Intergovernmental Oceanographic Commission of UNESCO (1985, 1994, 2001). Details of individual instruments for measuring sea level are normally available from the manufacturers. Comprehensive accounts of the principles and applications of satellite altimetry are given in Fu and Cazenave (2001).

Questions

1.1 In Figure 1.1, why do the periods of smaller tidal ranges (neap tides) occur earlier in successive months?

1.2 In Figure 1.3, why are the maximum length and time scales 40 000 km and 4.5 bn years?

1.3 Why is it wrong to use the uncorrected depths shown on nautical charts, in computer modelling of regional sea level variations?

1.4 Discuss the type of sea level instrumentation that might be suitable when making measurements:

(a) to provide a short-term record of levels of limited accuracy to support an offshore hydrographic survey;
(b) to assist with shipping movements in a harbour;
(c) for evaluating the risks of local flooding;
(d) for studies of long-term global sea level changes.

Chapter 2
Tidal forces and patterns

The main causes of sea level changes are the regular tidal movements caused by the gravitational forces of the moon and sun. Although observations of sea level in different parts of the world show an often bewildering diversity of tidal patterns and behaviour, there are common features in all these variations which follow from the basic tidal forces that produce them. In this chapter we will look in more detail at these forces and the patterns they produce. We will use basic mathematics to explain the principles, while reserving a more detailed mathematical development, for interested readers, for Appendix 1. This chapter also includes an explanation of the geoid surface, essential for the mean sea level discussions in Chapter 7.

Modern tidal theory began with, and remains founded upon, Newton's formulation of the Law of Gravitational Attraction: that two bodies attract each other with a force that is proportional to the product of their masses and inversely proportional to the square of the distance between them. Newton (1642–1727) was able to show why there are generally two tides a day. He also showed why the half-monthly spring to neap cycle in tidal range occurs; why once-daily diurnal tides are a maximum when the moon is furthest from the plane of the equator; and why the tides at the equinoxes in March and September are larger than those at the solstices in June and December.

2.1 Tidal diversity

As shown in Figure 1.1, the two main tidal features of any sea level record are the *range*, measured as the height between successive high

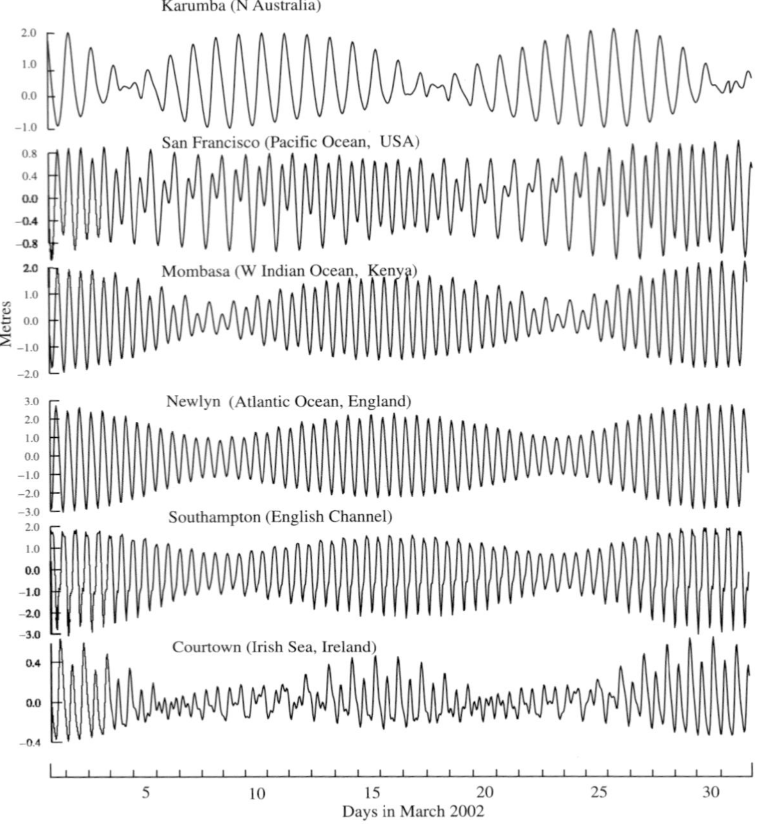

Figure 2.1. Tidal predictions (in metres) for March 2002 at six sites around the world have very different tidal regimes. At Karumba the tides are diurnal, at San Francisco they are mixed, whereas at both Mombasa and Newlyn the semidiurnal tides are dominant. The semidiurnal tides at Southampton and Courtown are strongly distorted by the influence of the shallow water.

and low levels, and the *period*, the time between one high (or low) water and the next high (or low) water. Both of these tidal features vary greatly from one site to another. Figure 2.1 shows the tides for March 2002 at six sites around the world. This month was selected because it shows large changes in tidal range and because the different astronomical effects can be clearly distinguished. Figure 2.2 shows the simultaneous changes in the moon's phase, distance and declination, to which the changing tides can be related. We can look at the relationships between the tides and astronomy by comparing the details of these two figures.

In most of the world's oceans the dominant regular tidal pattern is similar to that shown for Newlyn in the North Atlantic and for Mombasa on the African shore of the Indian Ocean. Each tidal cycle takes an average of 12 hours 25 minutes, so that two tidal cycles occur for each upper transit of the moon (every 24 hours 50 minutes). Because each tidal cycle occupies roughly half of a day, this type of tide is called semidiurnal. Semidiurnal tides have a range which typically increases and decreases cyclically over a fourteen-day period. The maximum ranges,

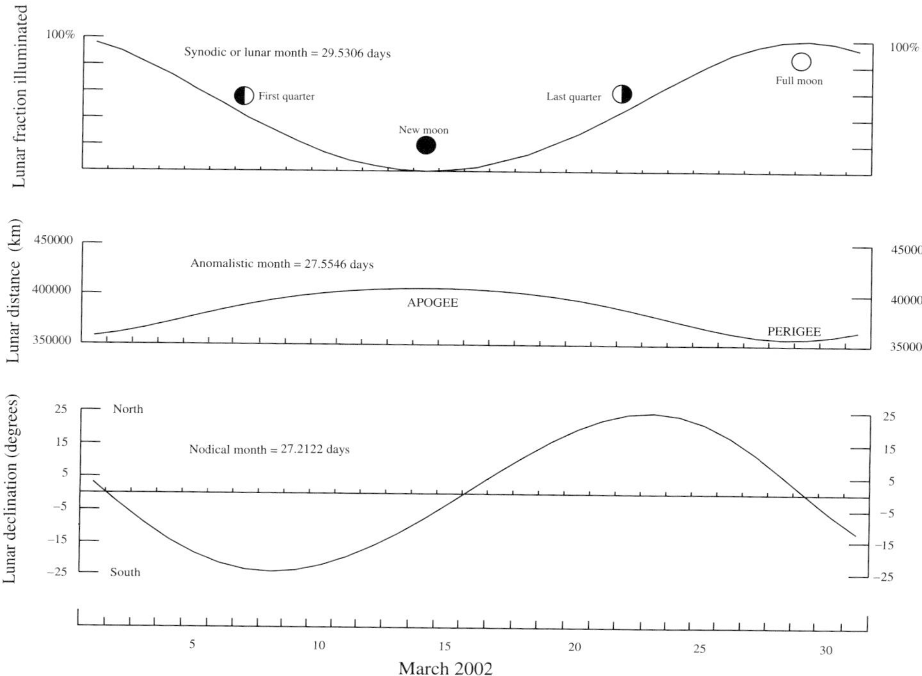

Figure 2.2. The lunar characteristics responsible for the tidal patterns in Figure 2.1. Solar and lunar tidal forces combine at new and full moon to give large spring tidal ranges. Lunar distances vary through perigee and apogee over a 27.55-day period. Lunar declination north and south of the equator varies over a 27.21-day period. Solar declination is zero on 21 March.

called spring tides, occur one or two days after both new and full moons (15 and 29 March 2002 in Figure 2.1). These times are termed *syzygy*: when the moon, earth and sun are in line. The minimum ranges, called neap tides, occur shortly after the times of the first and last quarters (7 and 22 March in Figure 2.1). The relationship between tidal ranges and the phase of the moon with a repeat cycle of 29.5 days is due to the additional tide-raising attraction of the sun, which reinforces the moon's tides at syzygy, but reduces them at the first and last quarters (quadrature).

Figure 2.2 also shows that when the moon is at its maximum distance from the earth (around 15 March 2002 in this example), known as lunar apogee, semidiurnal tidal ranges are less than when the moon is at its nearest approach (around 1 and 30 March 2002), known as lunar perigee. This variation in the moon's distance, from perigee through apogee and back to perigee, is repeated every 27.55 solar days. Big semidiurnal tidal ranges occur when spring tides (syzygy) coincide with lunar perigee, whereas small semidiurnal ranges occur when neap tides (quadrature) coincide with lunar apogee.

Semidiurnal tidal ranges increase and decrease at roughly the same time everywhere, but there are great variations in the size of these tidal ranges. The maximum semidiurnal tidal ranges are found in partly enclosed seas. In the Minas Basin in the Bay of Fundy (Canada), the semidiurnal tides at Burncoat Head have a mean spring range of 12.9 m. The mean spring ranges at Avonmouth in the Bristol Channel (United Kingdom) and at Granville in the Gulf of St Malo (France) are 12.3 m and 11.4 m respectively. In Argentina the Puerto Gallegos mean spring tidal range is 10.4 m; at Bhavnagar, north of Mumbai, India, it is 8.8 m; and the Korean port of Inchon has a mean spring range of 8.4 m. More generally, however, the ranges are much smaller. In the main oceans the semidiurnal mean spring tidal range is usually less than 2 m.

At certain times in the lunar month the semidiurnal high water levels are alternately higher and lower than the average. This behaviour is also observed for the low water levels, the differences being most pronounced when the moon's declination north and south of the equator is greatest. The differences can be accounted for by introducing a small extra tide with a period close to one day, which adds to one high water level but subtracts from the next one. Later we will develop the idea of a superposition of several partial tides, each of different periods, to produce the observed sea level variations at any particular location.

Returning to Figure 2.1, at San Francisco on the west coast of the USA, the tides with a one-day period, which are called diurnal tides, are similar in magnitude to the semidiurnal tides. This composite type of tidal regime is called a mixed tide, the relative importance of the semidiurnal and the diurnal components changing throughout the month (see also Figure 9.6c). The diurnal tides are most important when the moon's declination is greatest but reduce to zero when the moon is passing through the equatorial plane, when it has zero declination. Compare this with the spring–neap semidiurnal tidal changes which are only partly reduced during the period of neap tides.

In a few places the diurnal tides are much larger than the semidiurnal tides. Karumba in the Australian Gulf of Carpentaria is the example shown at the top of Figure 2.1. Here the tides reduce to zero when the moon's declination is zero, increasing to their largest values when the moon is at its greatest declination, either north or south of the equator. Large diurnal tides, although relatively uncommon, are also found in part of the Persian Gulf, the Gulf of Mexico, part of the South China Seas and the Sea of Okhotsk.

The tidal ranges on the relatively shallow continental shelves, for example at Newlyn and Southampton, are usually larger than those of the deep oceans that they surround. However, very small tidal ranges are observed in some shallow areas, often accompanied by curious distortions of the normal tidal patterns. In Figure 2.1 the curves for Courtown on the Irish coast of the Irish Sea, show that there the range varies from more than 1 m at spring tides to only a few centimetres during neap tides. When the range is very small, there are four tides a day.

These effects are due to the distorted tidal propagation in very shallow water. Shallow water distortions of the tides, which are discussed in detail in Chapter 5, are also responsible for the double high water feature at Southampton in the English Channel, where semidiurnal tides prevail. Double high waters are also found at Den Helder in the North Sea and at Le Havre in the English Channel. Double low waters occur along the Dutch coast of the North Sea from Haringvlietsluizen to Scheveningen, where they are particularly well developed at the Hook of Holland.

2.2 Gravitational attraction

The essential elements of a physical understanding of the tides are Newton's Law of Gravitational Attraction together with Newton's Laws of Motion, and the Principle of Conservation of Mass. It will be useful to remind ourselves about these basic physical constraints.

Newton's First Law of Motion asserts that a body in motion, which for our purposes can be an element of seawater, continues at a uniform speed in a straight line unless acted upon by a force. The Second Law relates the acceleration, the rate of change of motion or momentum, to the magnitude of the imposed force: the acceleration takes place in the direction of the force. For this to be valid, motion, for example in a straight line, must be observed in an external system of space coordinates, not on a rotating earth. Motion which appears to be in a straight line to an observer on a rotating earth follows a curved path when observed from space. This curvature must be produced by an additional terrestrial force, even though this force may not be immediately apparent. The effects of the earth's rotation on the propagation of tidal waves will be discussed in more detail in Chapter 4.

The Law of Gravitational Attraction states that two particles, of masses m_1 and m_2, separated by a distance r are mutually attracted by a force:

$$F = G\frac{m_1 m_2}{r^2} \tag{2.1}$$

G is the universal gravitational constant. Fortunately we do not have to consider the mass of every particle individually. The total gravitational attraction between two large masses such as the earth and the moon can be calculated by assuming that for each body the total mass is concentrated at a single point. For a sphere this point is at its centre.

Table 2.1. *The basic astronomical constants of the moon–earth–sun system, which controls the tidal forces and amplitudes.*

The moon			
Mass	m_l	7.35×10^{22}	kg
Mean radius		1738	km
Mean distance from the earth	r_l	384 400	km
		60.3	earth radii
The earth			
Mass	m_e	5.97×10^{24}	kg
		81.3	lunar masses
Equatorial radius	a	6378	km
Mean distance from the sun	r_s	149 600 000	km
		23 460	earth radii
Mean distance from centre of the earth to earth–moon mass centre		4671	km
The sun			
Mass	m_s	1.99×10^{30}	kg
		332 946	earth masses
Radius		696 000	km

Table 2.1 summarises the physical constants of the moon–earth–sun system. The scales and distances involved are hard to appreciate, and so a relative comparison is helpful. If a table-tennis ball represents the moon, the earth may be represented by a sphere with the same radius as a table-tennis bat 4 m distant; a sphere of 15 m diameter some 2 km away may represent the sun.

Consider for the moment only the moon–earth system. In this argument we are also not considering the rotations of the bodies about their own axes; the important influence of this rotation on the generation of tides will be introduced later. The two bodies will revolve about their common *centre of mass* with a period which is called the sidereal period. The sidereal period for the moon–earth system is 27.32 days, defined as one sidereal month. The centre of mass of two bodies lies on a line between them: for two bodies of equal mass it is in the middle. For the moon–earth system the centre of mass actually lies within the earth, because the earth is 81 times more massive than the moon. The necessary acceleration of each body towards this centre as they revolve about it is produced by their mutual gravitational attraction

The force necessary to give each particle of the earth the acceleration to perform this revolution is the same as the force for the particle at the

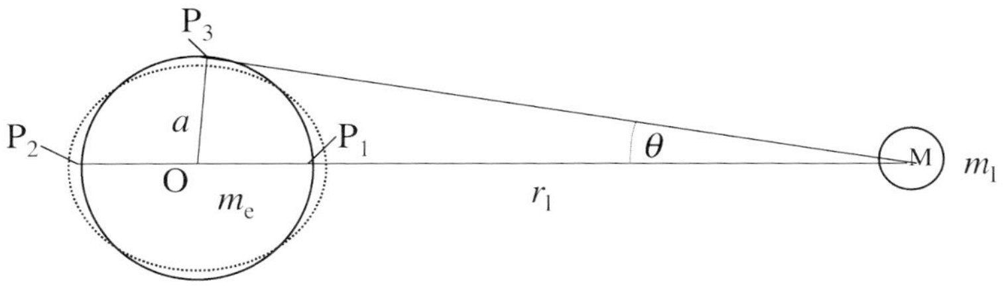

Figure 2.3. The positions in the earth–moon system that are used to derive the tidal forces. The separation is distorted but the relative diameters of the earth and moon are to scale.

centre of the earth: for each particle the gravitational force provides this necessary acceleration. For those particles nearer the moon than the centre particle, the gravitational attraction is stronger than is necessary to maintain the orbit. For those further away the forces are weaker. The differences between the forces necessary to maintain the particle's orbit, and the gravitational attractive forces actually experienced, generate the tides on the surface of the earth.

We can now look at what this means in more detail. Consider a particle of mass m located at P_1 in Figure 2.3, on the earth's surface. The centres of the earth and moon are separated by a distance r_1, and the earth radius is a. From Equation (2.1) the force towards the moon is:

$$G\frac{mm_1}{(r_1 - a)^2}$$

whereas the force necessary for its rotation is the same as for a particle at the centre of the earth's mass, O:

$$G\frac{mm_1}{r_1^2}$$

The difference between these is the tide-producing force at P_1:

$$Gmm_1\left[\frac{1}{(r_1 - a)^2} - \frac{1}{r_1^2}\right] = \frac{Gmm_1}{r_1^2}\left[\frac{1}{(1 - a/r_1)^2} - 1\right]$$

The first term within the square brackets can be expanded, using the mathematical approximation:

$$[1/(1 - \beta)^2] \approx 1 + 2\beta \text{ for small } \beta$$

For us β is $a/r_1 \approx 1/60$ and so the terms in the square brackets reduce to $2a/r_1$, to give a net force towards the moon of:

$$\text{Tidal force at P}_1 = \frac{2Gmm_1a}{r_1^3} \tag{2.2}$$

For a particle at P_2, the gravitational attraction is too weak to supply the necessary acceleration towards the centre of mass. As a result there is a net force away from the moon:

$$-\frac{2Gmm_1a}{r_1^3}$$

The net force at P_3 is directed towards the earth centre. For $r_1 \gg a$, the force along P_3M is:

$$\frac{Gmm_1}{r_1^2}$$

There is a small component of this force towards O that is given by:

$$\frac{Gmm_1}{r_1^2}\sin\theta \tag{2.3}$$

The strength of this component is found by making use of the approximation: $\sin\theta = \theta$ for small values of θ. In our case $\theta = a/r_1 \approx 1/60$. Hence the tidal force at P_3 is vertically towards the centre of the earth:

$$\frac{Gmm_1a}{r_1^3}$$

The net effect is for particles at both P_1 and P_2 to be displaced away from the centre of the earth, whereas particles at P_3 are displaced towards the centre. For a fluid earth this results in an equilibrium shape (assuming static conditions), which is slightly elongated along the axis between the centres of the moon and the earth and flattened across the poles. The dotted curve in Figure 2.3 shows this adjustment.

As drawn, Figure 2.3 shows the forces for the moon or sun overhead at the equator. However, if we now consider Figure 2.3 to show an equatorial section of the earth (that is, we are now looking down over one of the poles), we can introduce the rotation of the earth on its own axis (rotation about an axis perpendicular to the page, through O) while the equilibrium double bulge remains fixed. Each point on the circumference will pass through two maximum and two minimum levels for each daily rotation. This rotation results in the commonly observed two tides a day, called the semidiurnal tides.

These simple arguments have shown that the tide-producing forces depend on the finite radius of the earth a, the mass of the moon m_1, and the inverse cube of the lunar distance r_1.

We can simplify Equation (2.3) by replacing the inconvenient universal gravitational constant G by more accessible parameters. The vertical gravitational force on a particle of mass m on the earth's surface due to its attraction to the centre of mass of the earth is given by Equation (2.1):

$$mg = \frac{Gmm_e}{a^2} \tag{2.4}$$

where g is the local gravitational acceleration. So the tidal force at P_1 may be written:

$$2mg\left(\frac{m_l}{m_e}\right)\left(\frac{a}{r_l}\right)^3 \tag{2.5}$$

From the values in Table 2.1, this acceleration is approximately:

$$2g\left(\frac{1}{81.3}\right)\left(\frac{1}{60.3}\right)^3 = 11.2 \times 10^{-8}g$$

so that the value of g is very slightly reduced at P_1 and P_2. A person weighing 100 kg would weigh 11.2 mg less as they passed through these positions!

2.3 Tidal forces: a fuller description

These limited arguments in terms of forces can take us only so far. A more elegant approach in terms of gravitational potential is given in Appendix 1. Based on the fuller development, Figure 2.4 shows how the balance of gravitational and centrifugal forces results in a pattern of vertical and horizontal tidal forces over the surface of the earth.

The vertical forces (Figure 2.4a) produce small changes in the weight of a body, as previously discussed. However, it is the small horizontal forces (Figure 2.4b) that produce the accelerations necessary to invoke tidal water movements. It is possible to regard these variations in the horizontal tidal forces as due to small periodic rotary tilting of the horizontal plane at each point on the surface.

The tidal forces change as the coordinates of the sub-lunar point change over a day and a month (due to the earth's rotation and the changes in the lunar declination (d_l)) in terms of longitude and latitude; and as the moon's distance from the earth (r_l) increases and decreases in its elliptical orbit.

2.3.1 The Equilibrium Tide

Based on the full analysis we can compute an *Equilibrium Tide* that takes these changes into account. The Equilibrium Tide is defined as the elevation of the sea surface that would be in equilibrium with the tidal forces if the earth were covered with water and the response of the water to the tidal forces were instantaneous.

The Equilibrium Tide has three coefficients that characterise the three main *species* of lunar tides:

- the *long-period species*, with tidal changes over a month and longer; these are due to changes in the lunar distance (r_l) and declination (d_l);

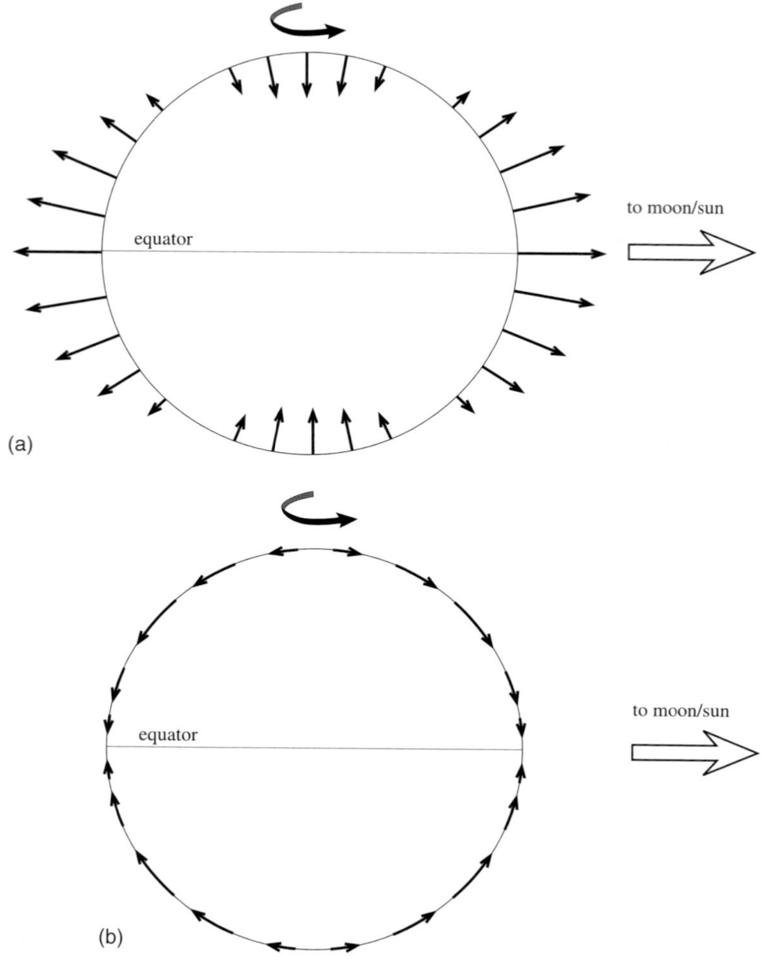

Figure 2.4. The distribution of tidal forces around the earth. (a) Vertical tidal forces, which are greatest at the equator, zero at 35° latitude and reversed at the poles. (b) Horizontal tidal forces; the maximum horizontal force is at 45° latitude.

- the *diurnal species* at a frequency of around one cycle per day, controlled by the lunar declination (d_l) and the earth's rotation (here we anticipate the explanation given in Section 2.4.1);
- and the *semidiurnal species* at two cycles per day, controlled by the earth's rotation.

There are important similarities and differences between the Equilibrium Tide and real tides. Although the Equilibrium Tide bears no spatial resemblance to the real observed ocean tide, its development is essential because it serves as an important reference system for tidal analysis. The amplitudes of the Equilibrium Tides are small. For the semidiurnal tide at the equator when the declination is zero, the amplitude calculated using the values in Table 2.1 is 0.27 m. The observed semidiurnal ocean tides are normally much larger than the Equilibrium Tide because of the dynamic response of the ocean to the tidal forces, as discussed in Chapter 4.

However, the observed tides do have their energy at the same frequencies as the Equilibrium Tide. For us the importance of the Equilibrium Tide lies in its use as a reference to which the observed phases and amplitudes of harmonic tidal constituents can be related, as described in Chapter 3. It also gives an indication of the most important constituents to be included in an efficient method for tidal analysis and prediction.

2.3.2 Solar tides

The tidal forces due to the sun are calculated in the same way that we have calculated the tidal forces due to the moon, by replacing m_l and r_l by m_s and r_s in Equation (2.5). Using the values in Table 2.1, the acceleration is:

$$2g(332\,946)\left(\frac{1}{23\,460}\right)^3 = 5.2 \times 10^{-8}g$$

The solar tidal forces are a factor of 0.46 weaker than the lunar tidal forces, because the much greater solar mass is slightly more than offset by its greater distance from the earth.

2.4 Tidal patterns

Detailed mathematical descriptions of the Equilibrium Tide and of the lunar and solar parameters are necessary for a rigorous development of the full pattern of variations in the tidal forces, but several features of the observed tides plotted in Figure 2.1 can be explained in more general terms. These features include the relationship between lunar and solar declination; large diurnal tides; and the spring–neap two-weekly cycle in the range of the semidiurnal tides.

2.4.1 Diurnal tides

If we compare the tidal changes of sea level plotted in Figure 2.1 for Karumba with the lunar changes plotted in Figure 2.2, we observe that maximum diurnal tidal ranges occur when the lunar declination is greatest, and that the ranges become very small when the declination is zero. This is because the effect of declination is to produce an asymmetry between the two high and the two low water levels observed as a point P

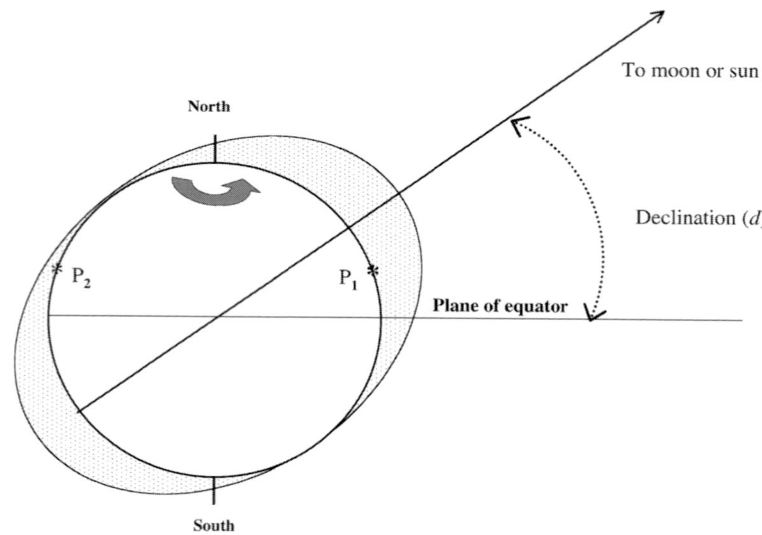

Figure 2.5. This diagram shows how unequal semidiurnal tides are produced when the moon or sun is north or south of the equator, causing generation of diurnal tides. The diurnal tides increase as the declination increases. For the moon the declination reaches a maximum every 18.6 years.

rotates on the earth within the two tidal bulges. In Figure 2.5, where these tidal bulges have been greatly exaggerated, the point at P_1 is experiencing a much higher Equilibrium Tide level than it will experience half a day later when the earth's rotation has brought it to P_2.

According to the theory of Equilibrium Tides, the two high water levels in one day would be equal if P were located on the equator. However, this is not observed in practice, as we shall discuss in Chapter 4, because the ocean responses to tidal forcing are much more complicated.

The lunar declination varies over a nodical month of 27.2 days. The solar declination varies seasonally from 23.5°N in June to 23.5°S in December, and back again. In Figure 2.1 the diurnal tides at Karumba approach zero amplitude when the lunar declination is zero because during the month of March the solar declination is also zero. Similarly, the total amplitude of the diurnal forcing becomes very small in September. In other months the solar diurnal forces are present even when the moon's declination is zero, and so the total diurnal tides will not completely disappear. The largest diurnal tides occur in June and December when the combined lunar and solar contribution is greatest.

At times and places where the solar diurnal tides are important there is an interesting seasonal variation. If the diurnal high water levels occur between midnight and noon during the summer, then they will occur between noon and midnight during the winter. This is because of the phase reversal of the equilibrium solar bulges (Figure 2.5) and hence the

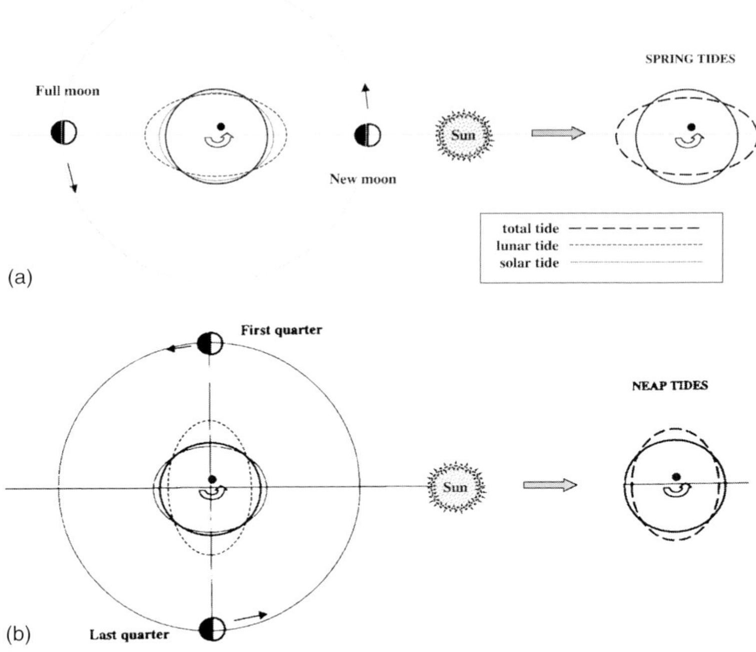

Figure 2.6. (a) Conditions for spring tides at new and full moon (syzygy). (b) Conditions for neap tides at quarter moons (also called lunar quadrature).

diurnal tidal forces. In the next chapter we shall represent this effect by a tidal term $\mathbf{K_1}$.

2.4.2 Spring–neap cycles

The fortnightly modulation in semidiurnal tidal amplitudes is due to the various combinations of the lunar and solar semidiurnal tides. At Mombasa, Newlyn, Southampton and Courtown in Figure 2.1, maximum ranges are seen shortly after the times of new moon and full moon. The minimum ranges occur at first quarter and last quarter (also called lunar quadrature). This is because at times of spring tides the lunar and solar forces combine together, but at neap tides the lunar and solar forces are out of phase and tend to cancel. Figure 2.6 shows how the phases of the moon and the ranges of the semidiurnal tides are related.

On average the Equilibrium spring tides (moon+sun) are 1.46 times bigger than the Equilibrium lunar tide, and the neap tides (moon−sun) are 0.54 less. As we shall see in Chapter 4, ocean responses to the tidal forces change these Equilibrium ratios, so the observed tides are somewhat different. In practice the observed spring and neap tides lag the maximum and minimum of the tidal forces, usually by one or two days (called the *age of the tide*), as the oceans need time to respond to the changing forces. The synodic period, from new moon to new moon, is 29.5 days, and the time from one spring tide to the next is 14.8 days.

Within a lunar synodic period the two sets of spring tides are usually of different amplitudes. We can examine Figure 2.1 in more detail. The Mombasa record shows that the ranges at the beginning and end of the month are larger than those in the middle. This difference is due to the varying lunar distance: lunar apogee (the furthest distance of the moon from the earth) occurs just after the new moon on 15 March 2002 (Figure 2.2), whereas the moon is near perigee (nearest to the earth) for the spring tides at the start and end of the month. For this reason the middle spring tides are smaller than those at the start and end of the month.

One complete cycle from perigee to perigee takes an anomalistic month of 27.6 days. At perigee the lunar tidal forces are 15 per cent greater than average. At apogee the lunar forces are 15 per cent less than average. These modulations in the total semidiurnal forces due to changes in the moon's distance are smaller than the 46 per cent modulations due to the sun, but their effects are significant for tidal ranges. Around the Atlantic coast of Canada the apogee–perigee effects are particularly influential because of favourable regional responses.

A third modulation of the semidiurnal forces is produced by the varying declinations of the moon and sun. As the strength of the diurnal forces increases the semidiurnal forces become weaker. Maximum semidiurnal forces occur when the moon and sun are both in the equatorial plane. As the moon and sun move north and south of the equator the diurnal forces increase, but the semidiurnal forces are less. For example, the semidiurnal lunar forces are reduced by 23 per cent when the moon has its maximum declination of 28.6°. The solar semidiurnal forces are reduced by 16 per cent in June and December when the declination reaches its maximum value of 23.5°. In March and September near the equinoxes, with the sun overhead at the equator, the solar semidiurnal forces are maximised with the result that spring tides near the equinoxes are larger than usual. These big tides are called equinoctial spring tides.

2.4.3 Nodal cycles

The earth's equatorial plane is inclined at 23° 27' to the plane in which the earth orbits the sun (called the ecliptic). This inclination causes the seasonal changes in our climate, and the regular seasonal movements of the sun north and south of the equator. The plane in which the moon orbits the earth is inclined at 5° 09' to the plane of the ecliptic; this plane rotates slowly over a period of 18.61 years. As a result, over this 18.61-year *nodal* period the amplitude of the lunar declination increases and decreases slowly. The maximum lunar monthly declination north and south of the equator varies between 18° 18' and 28° 36'. There are maximum values of lunar declination in 1969, 1987, 2006 and 2025, and minimum values in 1978 and 1997, 2015 and 2034.

Increases in the range of lunar declination over 18.6 years increases the amplitude of the diurnal lunar tides, but as explained at the end of Section 2.4.2, there is a corresponding decrease in the amplitude of the semidiurnal lunar tides. These nodal modulations decrease the average lunar semidiurnal Equilibrium Tide by 3.7 per cent when the declination amplitudes are greatest, with a corresponding 3.7 per cent increase 9.3 years later. The effects on observed sea level variations at Newlyn, where the tides are strongly semidiurnal, are shown in Figure 3.2.

2.5 The geoid

We end this chapter by introducing the *geoid*, a very important concept for mean sea level and its variations worldwide.

The undisturbed surface of the ocean, about which the tides oscillate, is not really spherical or even ellipsoidal in shape. If there were no tidal forcing, no differences in water density, no currents and no atmospheric forcing, the sea surface would adjust to take the form of an equipotential surface. An equipotential surface is one over which a particle can move without any vertical movement against the force of gravity. Its surface is everywhere horizontal and the vertical, by definition, is the direction at right angles to this surface. This undisturbed equipotential surface is called the *geoid*. Its exact shape depends on the distribution of mass within the earth, and on the rate of rotation of the earth about its own axis.

Over geological time the earth has adjusted its shape to the rotation by extending the equatorial radius and reducing the polar radius. The shape generated by this rotational distortion is called an ellipsoid of revolution (see Section 1.4.1).

If the earth consisted of matter of uniform density, then an ellipsoid would be an accurate description of the geoid; however, the actual geoid has positive and negative excursions of several tens of metres above and below a geometric ellipsoid, due to the uneven distribution of mass within the earth. This is shown in Figure 2.7. The geoid is more than 10 m higher than the ellipsoid at the North Pole, and 30 m lower than the ellipsoid at the South Pole. There is a geoid depression of −106 m corresponding to a mass deficiency south of India, and a geoid elevation of 73 m north

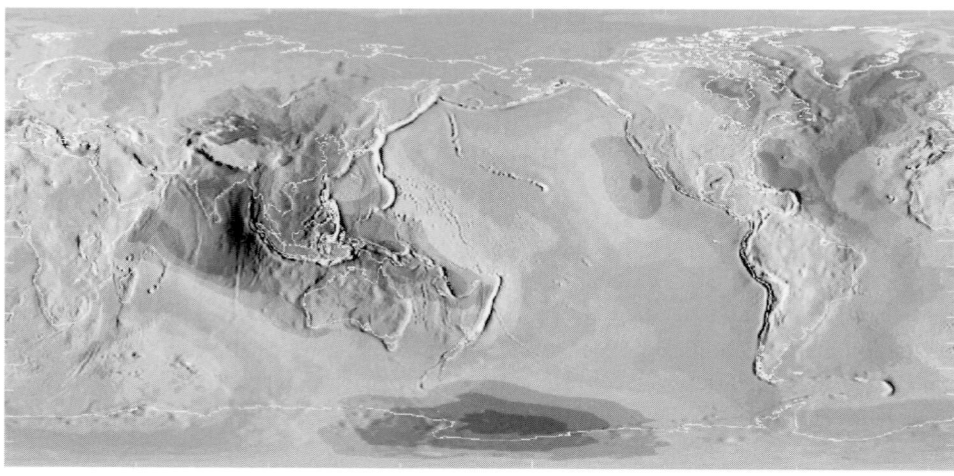

Figure 2.7. Geoid heights relative to the ellipsoid of revolution. These differences are due to uneven mass distribution in the earth. Minimum levels (−106 m, in purple) are found in the Indian Ocean. Maximum geoid levels (in orange) are about 85 m. The ocean trenches are clearly visible as a mass deficiency. This GEM 96 map was produced by the United States National Mapping Agency (NIMA) and NASA. See colour plate section.

of Australia, corresponding to mass excess. Interestingly, the highs and lows of the geoid show no direct correlation with the major tectonic plate surface features; they are thought to be due to density differences deep inside the earth. The mass deficiencies due to ocean trenches are easily seen.

Even local variations of mass distribution give local variations of several metres in the geoid and hence the mean sea level surface. These can be detected by satellite altimetry. One example of local geoid distortions often seen in altimeter records is due to submerged seamounts: the additional mass of a submerged seamount causes the geoid to rise. Over the Louisville ridge in the Pacific Ocean there is as much as 4 m of rise and fall of the geoid over a horizontal distance of 100 km. These local geoid changes are seen at the sea surface as local mean sea level anomalies; when detected by satellite altimeters they have been used to locate seamounts in uncharted ocean areas (see Figure 2.8).

Typical geoid slopes are around 1 in 10^{-5}, or 2 cm km^{-1}. To avoid cross-track geoid differences appearing as changes in sea level, repeat satellite orbits (see Section 1.4.4) must be exact to better than 1 km. In some extreme cases, for example near ocean trenches, continental margins, islands and seamounts, geoid slopes can be an order of

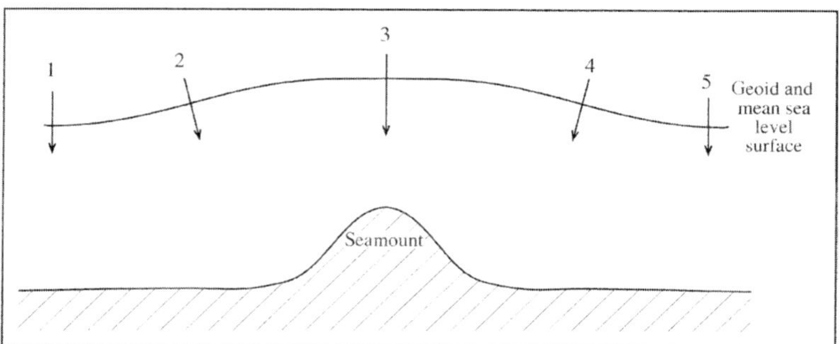

Figure 2.8. Local mass distribution changes the geoid. Here the gravitational attraction of the submerged seamount tilts the horizontal and raises the geoid above it. The mass deficiency of ocean trenches depresses the geoid. The size of these geoid excursions is typically a few metres.

magnitude greater, at 20 cm km^{-1}. As geological processes (e.g. isostatic adjustment after glacial melting) redistribute the mass of the earth, the shape of the geoid and the sea level surface gradually change. Some examples of these very slow geoid variations and sea level adjustments will be discussed in Chapter 7.

The geoid is an important reference surface for mean sea level studies. In the ideal conditions outlined at the start of this section mean sea level would follow the geoid exactly. However, in the same way as the geoid varies by tens of metres from the ellipsoid of revolution, the mean sea level surface around the globe varies up and down from the geoid by as much as a metre (see Figure 7.4). This is because in reality there are other forces that cause the mean sea level surface to deviate from the geoid surface: these include air pressure and winds; ocean density changes; and ocean currents. Because by definition the force of gravity is always perpendicular to the geoid surface, there are no horizontal gravitational forces along the geoid surface to disturb the water. For this reason, physical oceanographers need to know sea levels relative to the geoid in order to calculate the horizontal pressure gradient forces acting on and moving the water. This requirement calls for exact measurements of the shape of the geoid.

Special dedicated gravity missions are essential for improving our knowledge of the geoid to the necessary accuracy. These missions include the German gravity and magnetic satellite mission CHAMP (2000), and the Gravity Recovery and Climate Experiment (GRACE), which uses a pair of low earth-orbiting satellites whose relative distance is determined with high accuracy. Another mission is the Gravity Field and Steady-state Ocean Circulation Explorer (GOCE), based on a three-axis gravity gradiometer, with GPS precise orbit tracking in a very low orbit. GOCE plans to define the geoid undulations to within 1 cm over distances of 100 km.

Extreme tidal forces

By simple arguments it is evident that extreme tidal forces will occur when the moon and sun are in line with the earth and at their closest separations. At present, perihelion (the nearest approach of the earth to the sun) occurs early in January. For maximum semidiurnal tides, the moon and sun should also have zero declination. The condition of simultaneous perihelion and zero solar declination will next occur in March 6581.

However, there are several times when zero solar declination (the equinox) and lunar perigee occur almost simultaneously with zero lunar declination. Times of extreme semidiurnal tidal forcing are given in Table 2.2 for the period 1980–2028, together with the corresponding values of the orbital parameters for distance and for declination d. These times have been identified by predicting the hourly values of the Equilibrium semidiurnal tide (moon and sun) at the Greenwich meridian for the months of March and April, and of September and October, in each year. The highest value in each period was then tabulated, provided that it was greater than 0.4540 m. For comparison, the Equilibrium average semidiurnal lunar tidal amplitude is 0.2435 m. The highest value in the period (8 March 1993) is 1.89 times greater than the average lunar tide. Exact coincidence of other conditions with the solar equinox is not essential for very big tides, as is illustrated by the values in early March in 1993 and 1997, and October 2002.

Analyses of predicted extreme equinoctial spring tidal ranges over several years show a cycle with maximum ranges occurring approximately every 4.5 years. These maximum ranges occur when the moon is closest to the earth in its monthly orbit at either the March or September equinox, during an 8.85 -year cycle. Other periodicities in the recurrence of extreme tides can be expected as different factors in the relationship combine; for example, lunar perigee and zero lunar declination coincide every 6 years.

Highest and lowest astronomical semidiurnal tides, published in official tables, and sometimes used as datum levels, are often estimated by looking through several years of tidal predictions. Although maximum predicted tidal ranges usually result from extremes in the Equilibrium tidal forcing, local ocean responses to these forces may produce local extreme tides at times other than those tabulated. For maximum diurnal tides a different set of conditions would apply.

Of course, the small differences between the extreme tidal predictions in Table 2.2 are of mainly academic interest as the actual extremes will also depend on the prevailing weather conditions discussed in Chapter 6.

Table 2.2. *Times of maximum semidiurnal tidal forces during the period 1980–2029 and the corresponding astronomical arguments. Only the highest values in each equinoctial period are given.*

				Moon		Sun	
Year	Day		Equilibrium amplitude (m)	Distance ratio to mean	$\cos d_l$	Distance ratio to mean	$\cos d_s$
1980	16	Mar.	**0.4575**	1.0748	0.992	1.0051	1.000
1984	25	Sept.	**0.4544**	1.0768	0.998	0.9972	1.000
1993	8	Mar.	**0.4597**	1.0778	0.999	1.0075	0.996
1997	9	Mar.	**0.4550**	1.0733	0.998	1.0072	0.997
1997	17	Sept.	**0.4553**	1.0765	1.000	0.9950	0.999
1998	28	Mar.	**0.4588**	1.0764	1.000	1.0020	0.999
2002	27	Mar.	**0.4570**	1.0764	0.994	1.0020	0.999
2002	4	Oct.	**0.4566**	1.0761	0.999	1.0001	0.996
2007	18	Mar.	**0.4552**	1.0725	1.000	1.0046	1.000
2011	18	Mar.	**0.4575**	1.0761	1.000	1.0047	1.000
2015	19	Mar.	**0.4554**	1.0749	1.000	1.0043	1.000
2015	27	Sept.	**0.4544**	1.0771	1.000	0.9978	1.000
2020	8	Apr.	**0.4549**	1.0768	0.999	0.9987	0.992
2028	11	Mar.	**0.4559**	1.0737	1.000	1.0067	0.998
2028	18	Sept.	**0.4540**	1.0755	0.999	0.9952	1.000

Planetary tidal forces

There are also tidal movements in the solid earth in response to the tidal forces. These solid earth movements are smaller than the Equilibrium ocean tide, and the response to the gravitational forces is almost instantaneous. Earth tides are of interest for studying the elastic properties of the earth, and must be taken into account in modelling of the ocean response to tidal forcing.

The other planets in the solar system produce negligible tidal forces on earth. According to Equation (2.5), Venus, whose mass is 0.82 m_e, and whose nearest approach to the earth is approximately 6500 earth radii, will give a maximum tidal acceleration of:

$$2g(0.82)\left(\frac{1}{6500}\right)^3 = 6.0 \times 10^{-12}g$$

which is only (0.000 054) of the moon's tidal acceleration. Although Jupiter has a mass of $318m_e$ the maximum tidal accelerations which it produces on earth are an order of magnitude less than those of Venus

(0.000 006) because of Jupiter's greater distance from the earth. Mars (0.000 001), Mercury (0.000 000 3) and Saturn (0.000 000 2) have even smaller effects.

The tidal effects of the earth on the moon produce a small elongation along the earth–moon axis, but because the moon always presents the same face to the earth, there are no tides on the moon equivalent to the diurnal and semidiurnal tides on earth. Some flexing of the solid moon occurs as the earth–moon distance changes in its elliptical orbit. There are also small changes due to a slight rocking motion in longitude in the lunar orbit, called lunar librations.

The satellites of Jupiter suffer much more dramatic tidal effects. Io has a similar mass to our moon and its orbital distance from Jupiter is almost the same as our moon is from the earth. However, because Jupiter has such an enormous mass, the Equilibrium Tides on the solid surface of Io can be up to 100 m high. The regular flexing, every 42 hours, heats the interior of Io, energising several active volcanoes. Europa, also a satellite of Jupiter, has weaker tides and a thick covering of ice. Cracks as long as 1500 km in the ice surface may be due to tidal stresses and the orbits of the other satellites. If, as many scientists suspect, there is a thick layer of melted water below its ice cover, then the tides on Europa may be the nearest equivalent in the solar system to our own ocean tides.

Further reading

The relative movements of the moon, earth and sun are analysed in detail in several astronomical textbooks. Roy (1988) gives a thorough treatment of orbital motion. Many of the references in Chapter 1 also cover tidal forces in varying degrees of detail.

Questions

2.1 At which times of the year will diurnal tides have their greatest amplitudes?

2.2 In Figure 2.1 the vertical axes are plotted on different scales. At which location are the semidiurnal tides greatest?

2.3 In Figure 2.2, what are the maximum lunar declinations? How does this change over an 18.6-year period?

2.4 From Equation (2.5) explain how the tidal forces on the surface of the earth would change if:

(a) the lunar distance doubled;
(b) the mass of the moon doubled;

(c) the radius of the earth were increased by 1 per cent but the radius of the moon remained the same.

Assume for illustration that all other factors remain the same in each case. Why is this assumption not valid?

2.5 A ships sails from a port in Sri Lanka, where there is a geoid depression of 100 m to a port in Europe where there is a geoid elevation of 50 m. How much work is done against gravitational forces in moving between these two levels? Is the ship further from the centre of the earth at the end of its voyage?

2.6 Use the periods for an anomalistic month and a nodical month given in Figure 2.2 to confirm the period of 6 years referred to in the text box.

Chapter 3
Analysis and prediction

We can now consider ways of extracting tidal information from sea level records, based on our understanding of the tidal forces developed in the previous chapter. We will also outline how regular horizontal water movements, tidal currents and streams can be analysed for tides using similar techniques. Finally, we show how the results of tidal analyses can be used to predict future tides from astronomical information.

Tidal analysis of data collected by observations of sea levels and currents serves two main purposes. Firstly, a good analysis provides the basis for predicting tides at future times, a valuable aid for shipping and other coastal operations. Secondly, the results of an analysis can be interpreted scientifically to extend our understanding of the hydrodynamics of the seas and their responses to tidal forcing.

We can define *tides* as periodic movements that are directly related in amplitude and phase to some periodic geophysical force. It is important to make a distinction between the popular use of the word "tide" to signify any change of sea level (this popular usage includes the term *tide gauge* for sea level measuring instruments), and our more specific use of the word to mean only the regular, periodic variations. These periodic movements are most apparent in the regular movements of the sea, but there are also tidal movements in the atmosphere and in the solid earth. As we showed in the previous chapter, the dominant periodic geophysical forcing is the variation of the gravitational field on the surface of the earth, caused by the regular movements of the moon–earth and earth–sun systems. Movements due to these astronomical gravitational forces are called *gravitational* tides. There are also much smaller movements due to regular meteorological forces: these are called either *meteorological* or, more usually, *radiational* tides.

In tidal analysis the aim is to produce significant time-stable tidal parameters, which describe the tidal regime at the place of observation. These parameters are often termed *tidal constants* on the assumption that the responses of the oceans and seas to tidal forces do not change with time. A good system of analysis represents the data well with a few significant stable numbers. A poor analysis represents the data by a large number of parameters which cannot be related to a physical reality, and for which the values obtained depend on the time at which the observations were made. If possible, an analysis should also give some idea of the confidence that can be attributed to each tidal parameter determined.

Here we will consider three basic methods specifically developed for tidal analysis. But before doing so we will show how some standard techniques for general time-series analysis can be used to extract important parameters to describe sea level variations. The first tidal analysis technique, which is now generally of only historical interest, is called the *non-harmonic* method, and relates high and low water times and heights directly to the phases of the moon and other coordinates of the moon and sun. The second method, which is the one most commonly used for predictions and for scientific work, *harmonic analysis*, treats the observed tides as the sum of a finite number of harmonic constituents with periods or angular speeds determined from the astronomical arguments. The third method develops the concept, widely used in electronic and mechanical engineering, of a frequency-dependent *response* of a system to a driving mechanism. In our case the driving mechanism is the tidal potential and the response is the observed ocean tide. The stilling well discussed in Chapter 1 is an example of a frequency-dependent system with a very small response to wind wave frequencies but a full response to the slower tidal changes of sea level.

3.1 Non-harmonic methods

The close relationship between the movements of the moon and sun and the observed tides makes the lunar and solar coordinates a natural starting point for any tidal analysis scheme.

The simplest and oldest technique of tidal analysis is to relate the time of local semidiurnal high water to the time of lunar transit through the local meridian or longitude. The time interval between lunar transit at new or full moon and the next high tide used to be known as the *local establishment*. Although this changes a little during a spring–neap cycle, the local establishment gives a very simple way of predicting times of high and low tides.

The *age of the tide* is an old but still useful term applied to the interval between the time of new or full moon and the time of maximum semidiurnal spring range, which usually occurs one or two days later.

The advantage of these non-harmonic parameters is their ease of determination and application. However, they are not powerful enough for a full tidal description and prediction scheme.

3.2 Basic statistics

Simple parameters that describe sea level changes are very valuable for a first scrutiny of our data. Before proceeding to more detailed tidal analyses, in this section we describe the simplest way of analysing a series of readings of sea level, and show how general methods of time-series analysis, available in standard computer spreadsheet software, can help us to classify the sea level changes into different periods (or frequencies).

One of the most fundamental statistics is the average of a series of sea level readings. Average sea levels, derived by analysis of long periods of sea level variations are used to define reference levels. These levels are often used for map or chart making or as a reference for subsequent sea level measurements. For geodetic surveys the *mean sea level* (MSL) is frequently adopted, being the average value of levels observed each hour over a period of at least a year, and preferably over about 19 years.

The general representation of the observed level $X(t)$ that varies with time (t) may be written:

$$X(t) = Z_0(t) + T(t) + S(t) \tag{3.1}$$

where $Z_0(t)$ is the mean sea level, which changes very slowly with time; we shall return to the slow changes in Chapter 7. The changes of Z_0 can be ignored for the rest of this chapter. $T(t)$ is the tidal part of the variation. $S(t)$ is considered to be the meteorological surge component. This surge component may also include other effects such as seiches, or rarely tsunamis (see Sections 6.6 and 6.7). Strictly, in any series of observations $S(t)$ will also include any errors in the measurements.

Measurements of sea levels and currents are conveniently tabulated as a series of hourly values, which for satisfactory analysis should extend over about a lunar month of 709 hours, or better over an average year of 8766 hours. Suppose that we have K observations of the variable $X(t)$ represented by $x_1, x_2, x_3 \ldots x_K$, then the mean is given by:

$$\overline{x} = \frac{1}{K}(x_1 + x_2 + x_3 + \cdots + x_K)$$

To describe the spread of the readings in a series of sea level measurements about the mean, we can compute the average variance σ^2, as the mean of the squares of the individual deviations:

$$\sigma^2 = \frac{1}{K}\sum_{k=1}^{K}(x_k - \overline{x})^2$$

which must always have a positive value. The square root of the variance σ is called the standard deviation of the distribution of x about $\overline{x}$.

A more powerful technique, Fourier analysis, represents a time series in terms of the distribution of its variance at different frequencies. Tidal and surge components of sea level series are usually independent because they have very different physical causes. It can be shown that, as a result of this independence, for summing over K sea level values the following relationship holds:

$$\sum_{k=1}^{K}(X(t_k) - Z_0)^2 = \sum_{k=1}^{K} T^2(t_k) + \sum_{k=1}^{K} S^2(t_k)$$

$$\text{Total variance} = \text{Tidal variance} + \text{Surge variance}$$

Here Z_0 is the mean sea level over the period of observation. We can divide both sides of the equation by K to get the average variance in each term.

It can also be shown that the variance is a measure of the energy in the sea level variations. This is an extension of the fact that the energy in a regular harmonic wind wave is related to the square of its amplitude. Information about the distribution of the energy in the sea level variations at different frequencies is of great interest in describing a series of observations. The sea level regime at a particular location may be characterised by the way in which the variance or energy is distributed among the different tidal species as discussed in Chapter 2 (diurnal, semidiurnal) and the non-tidal meteorological residuals (Chapter 6).

Table 3.1 shows the distribution of variance or energy for some sites with different sea level regimes, including the sites plotted in Figure 2.1. The first three columns show the variance or energy in the three main tidal species including the small amount in the long-period tides. The maximum diurnal variance is at Karumba, and the highest semidiurnal variance is at Newlyn. The fourth column shows the shallow-water tides, which are especially important at Newlyn and Southampton due to local effects. We will return to shallow-water tides in Chapter 5. The meteorological non-tidal effects are greatest at Buenos Aires. This breaking down of the sea level variability into component parts gives much more information than the other two cruder measures included in Table 3.1. These are the form factor and the macro/meso/micro categories in the final column (see Section 3.3.4).

Table 3.1. *The distribution of variance at eight sea level stations, showing how the energy is distributed in the different tidal species. Units are* cm^2.

	Tidal			Non-tidal					
	Long-period	Diurnal	Semi-diurnal	Shallow water	Meteorological	TOTAL	Form factor	Description	
Karumba	10	7000	1600	10	20	**71640**	7.67	Tropical extensive bay in northern Australia	Macrotidal
Honolulu	9	154	157	0	35	**355**	1.05	Ocean island surrounded by deep water	Microtidal
Mombasa	5	245	7555	2	19	**7826**	0.20	East African estuary near deep water	Mesotidal
San Francisco	10	240	6500	10	20	**6780**	0.85	Deep bay with open ocean access	Microtidal
Buenos Aires	20	160	440	20	2500	**3140**	0.69	Extensive estuary of River Plate, Argentina	Microtidal
Newlyn	17	37	17000	100	200	**17400**	0.05	Shallow site on northwest European shelf	Mesotidal
Southampton	120	50	10000	900	320	**11400**	0.07	Shallow site with strong non-linear interactions	Mesotidal
Courtown	116	55	284	55	222	**732**	0.55	Shallow site near Irish Sea amphidrome	Microtidal

However, although we can learn a lot about sea level records and describe their characteristics using the general methods outlined above, more detailed and specialised analyses of a different kind are required for tidal prediction. We can now look in detail at some of these specific tidal analysis techniques.

3.3 Harmonic analysis

3.3.1 Basic concepts

Even though individual tidal curves at individual coastal sites are very different, Figure 2.1 shows that the various locations considered all have tides that appear as periodic oscillations. It is natural for a detailed analytical system to take this regular waveform as a starting point. Mathematically, periodic oscillations are described in terms of an amplitude and a period or frequency. The basis of harmonic analysis is the assumption that tidal variations can be represented by a number N of harmonic terms of the form:

$$H_n \cos(\omega_n t - g_n) \tag{3.2}$$

where H_n is the amplitude; ω_n is the angular speed, which is related to the period T_n by $T_n = 2\pi/\omega_n$; g_n is the phase lag relative to some defined time zero; and t is time. The time zero for g_n is often taken as the phase lag on the Equilibrium Tide phase at the Greenwich Meridian, in which case it is called G_n. This use of the Equilibrium Tide as a reference for tidal analysis is one of its important functions. For a single harmonic, the difference between a high water level and the next low water level, the *tidal range*, is twice the amplitude ($2H_n$).

We can also use the Equilibrium Tide to determine the angular speeds ω_n of the constituents. These are found by an expansion of the Equilibrium Tide into harmonic terms: the speeds of these terms are found to have the general form:

$$\omega_n = i_a\omega_1 + i_b\omega_2 + i_c\omega_3 + (\omega_4, \omega_5, \omega_6 \text{ terms}) \tag{3.3}$$

where the values of ω_1 to ω_6 are the angular speeds related to astronomical parameters (see Table 3.2) and the coefficients i_a to i_c are small integers, usually in the range -2 to $+2$.

This abstract representation of the tides can seem a little remote from the observed movements of the moon and sun, but it is possible to relate individual tidal harmonics to real astronomical behaviour. Firstly, we can look at the simplest possible orbit of the moon and see how the tides this produces can be represented by a single harmonic term. Suppose the moon were to revolve continuously in an equatorial circle around the

Table 3.2. *The basic speeds and origin of the astronomical arguments (ω_n) that give the frequencies of the harmonic constituents. Note that* $\omega_0 = \omega_1 + \omega_2 - \omega_3$.

	Period	Degrees per mean solar hour	Symbol
Mean solar day	1.0000 msd	15.0000	ω_0
Mean lunar day	1.0351 msd	14.4921	ω_1
Sidereal month	27.3217 msd	0.5490	ω_2
Tropical year	365.2422 msd	0.0411	ω_3
Moon's perigee	8.85 years	0.0046	ω_4
Regression of moon's nodes	18.61 years	0.0022	ω_5
Perihelion	20942 years	—	ω_6

earth at a constant distance, with the earth rotating underneath it every 24 hours. The tides on the earth would have two maximum values (high waters) for each rotation period. However, because the moon will have moved on slightly in its monthly orbit (both earth and moon rotation are in the same sense), the two tides take 24 hours and 52 minutes. Each individual tide will follow the previous one at intervals of 12 hours 25 minutes and each high water will have the same amplitude.

The single harmonic, which represents these tides, is called $\mathbf{M_2}$. The naming convention is that M represents the moon and the subscript '2' shows that it is in the twice-daily semidiurnal tidal species. In this book we print tidal constituents in bold, to show that as vectors they have two variables associated with them – amplitude (H_n) and phase (g_n).

If the sun is also on the plane of the equator, the same principles apply to give a semidiurnal tide. This has a period of exactly 12 hours, because we base our clocks on the solar time. These two solar tides a day are represented by the symbol $\mathbf{S_2}$ using the same convention. We now have two tidal constituents. The combination of these, $\mathbf{M_2}$ and $\mathbf{S_2}$ will produce the spring–neap tidal cycle. In Figure 3.6 the same synthesis is presented as part of the tidal prediction method.

In the simple case of both the sun and moon moving in circular equatorial orbits relative to the earth, the compound tide would be completely represented by four parameters – the times of lunar and solar high waters and the amplitudes of the two tides. In our new system these are the amplitudes and phases of the harmonic tidal constituents $\mathbf{M_2}$ and $\mathbf{S_2}$. The next stage is to see how the concept of harmonic constituents can be developed to cover the real movements of the moon and sun.

The real movements of the moon and sun, north and south of the equator and their varying distances and angular speeds, can be considered tidally as the combined effects of a series of phantom satellites. These phantom satellites are of various masses and they move in various planes and at various speeds. Each satellite therefore has a simple tide, which is represented by an amplitude, and a time of high water. This concept formed the basis for the early development of harmonic analysis. In practice we can expand the astronomical expressions for declination (d_l and d_s) and distance (r_l and r_s) mathematically to determine the periods and theoretical amplitudes of the extra terms.

The first step in the development of harmonic analysis is to expand the Equilibrium Tide into a series of harmonic terms from the full expressions for the distance, declination and hour angle of the moon or sun. Including two extra harmonic terms at frequencies close to $\mathbf{M_2}$, which add to and subtract from $\mathbf{M_2}$ at different times can represent the varying moon–earth distance over a month. These two terms have angular speeds of 28.440° per hour and 29.529° per hour and relative amplitudes of 19.2 per cent and 2.8 per cent of the $\mathbf{M_2}$ amplitude. To emphasise their speed symmetry about the speed of $\mathbf{M_2}$ (28.984° per hour), these new harmonics are called $\mathbf{N_2}$ and $\mathbf{L_2}$ in the standard tidal notation. When the full expansion is done for both the moon and the sun the list of harmonic constituents is very long. A full list would include several hundred constituents.

Nevertheless, examination of the relative amplitudes of the constituents in the mathematical expansion of the Equilibrium Tide shows that in practice a few harmonics are dominant. These are usually $\mathbf{M_2}$, $\mathbf{S_2}$ and, to a lesser extent, $\mathbf{N_2}$. Similar expansions for the long-period and diurnal harmonic terms show that these species are also dominated by a few harmonics: for the long-period tides the annual term $\mathbf{S_a}$, which is enhanced by solar heating; and for the daily tides, usually $\mathbf{K_1}$ and $\mathbf{O_1}$. Table 3.3 lists the most important of these constituents, with the corresponding periods and angular speeds. The amplitudes in Table 3.3 are all given relative to the Equilibrium Tide amplitude $H_{M_2} = 1.0000$.

The line spectrum of tidal harmonic constituents for the diurnal and semidiurnal species are plotted in Figure 3.1. This shows the frequencies of the terms in a fuller expansion of the Equilibrium Tide, and confirms the importance of a few major tidal harmonics. The vertical scale is logarithmic, but the dominant terms are clearly seen.

The pattern of constituents depending on their period or frequency in Figure 3.1 can be explained in terms of Equation (3.3). The main divisions are the number of cycles per day (governed by i_a), and each of these is called a tidal *species*. The energy in these species is modulated by harmonic terms involving $i_b.\omega_2$, $i_c.\omega_3$, etc. In the complete expansion i_b, which fits the monthly modulations, varies from -5 to $+5$ and defines the *group* within each species. Within each group the value of i_c fits

Table 3.3. *The principal constituents of the tides and their physical causes.*

	i_b	i_c	Period (msd)	Speed (°per hour)	Relative coefficient ($M_2 = 1.0000$)	Origin	
Long-period $i_a = 0$							
$\mathbf{S_a}$	0	1	364.96	0.0411	0.0127	Solar annual	Strongly enhanced by seasonal climate effects
$\mathbf{S_{sa}}$	0	2	182.70	0.0821	0.0802	Solar semi-annual	
Diurnal $i_a = 1$							
$\mathbf{O_1}$	−1	0	1.076	13.9430	0.4151	Principal lunar	
$\mathbf{P_1}$	1	−2	1.003	14.9589	0.1932	Principal solar	
$\mathbf{K_1}$	1	0	0.997	15.0411	0.5838	Principal lunar and solar	
Semidiurnal $i_a = 2$							
$\mathbf{N_2}$	−1	0	0.527	28.4397	0.1915	Lunar ellipse	
$\mathbf{M_2}$	0	0	0.518	28.9841	1.0000	Principal lunar	
$\mathbf{L_2}$	1	0	0.508	29.5285	0.0238		
$\mathbf{S_2}$	2	−2	0.500	30.0000	0.4652	Principal solar	
$\mathbf{K_2}$	2	0	0.499	30.0821	0.1266	Declinational lunar and solar	

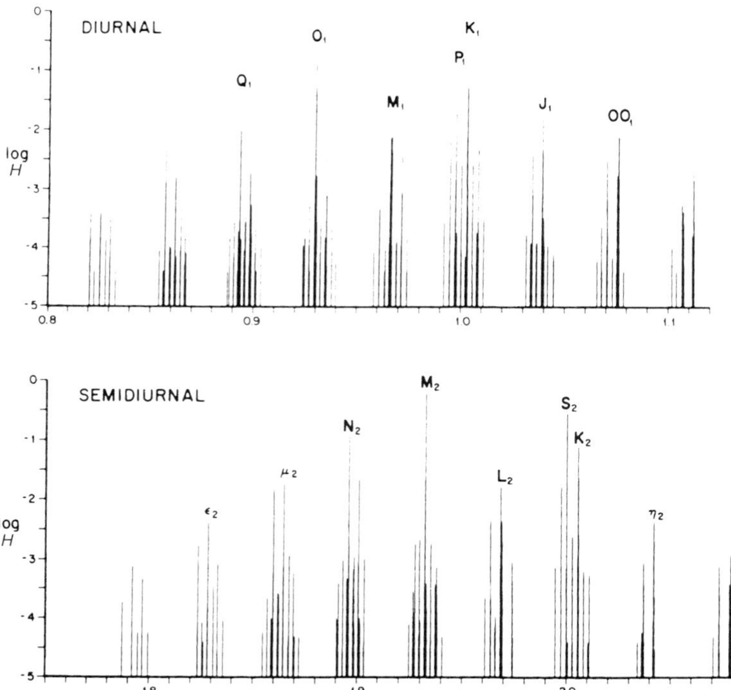

Figure 3.1. The Equilibrium Tide consists of a series of partial tides at discrete periods or frequencies. This figure shows the pattern of the periods found in the diurnal and semidiurnal tidal species, based on the expansion of Cartwright and Edden (1973). Each individual line represents a tidal constituent. Note the clustering of *constituents* into *groups* within each *species*. The height of the lines indicates the relative amplitude of the constituents, but the scale, plotted on a logarithmic axis, is compressed.

the annual modulations; it also varies from -5 to $+5$ and is said to define the *constituent*. In general, analysis of a month of data can only determine independently those terms which are separated by at least one unit in (ω_2) 0.5490° per hour. Similarly, analysis of a year of data can determine harmonic constituents which differ by one unit in (ω_3), 0.0411° per hour.

All the lunar constituents are affected by the *18.6-year nodal cycle*. Certain lunar constituents, notably $\mathbf{L_2}$, are also affected by an *8.85-year cycle* of the moon's perigee. These slower modulations, which cannot be independently determined from a year of data, must also be represented in some way. In the full harmonic expansion they appear as terms separated from the main term by the angular speeds $i_d\omega_4$ and $i_e\omega_5$. The terms that are separated by $i_f\omega_6$ (perihelion) may be considered constant for all practical purposes.

The modulations in ω_5 and ω_6, which cannot be resolved as independent harmonics by analysis of a year of data, are represented in harmonic expansions by small adjustment factors f and u to the amplitude H_n and the phase g_n. Each constituent is written in terms of degrees:

$$H_n f_n \cos[\omega_n t - (g_n + u_n)] \tag{3.4}$$

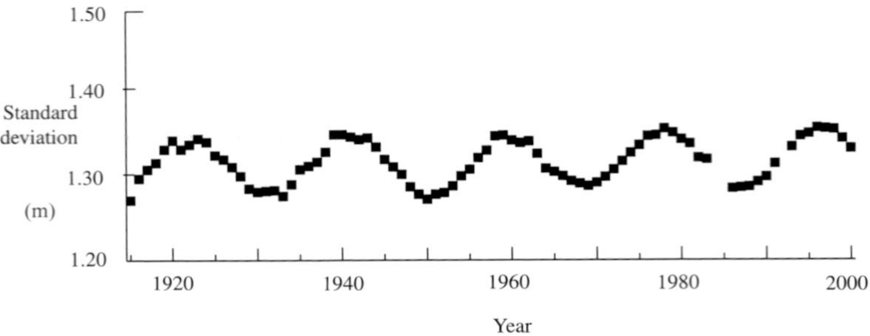

Figure 3.2. The amplitude of the semidiurnal lunar tide varies over an 18.6-year period due to changes in the amplitude of the lunar declination. This figure shows the standard deviation in the observed sea level variations at Newlyn: the 18.6-year modulations are evident. The semidiurnal tides are greatest when the declination range is least.

The nodal terms are f_n the nodal factor and u_n the nodal angle. The nodal factor and the nodal angle are 1.0 and 0.0 for the solar constituents as there are no nodal effects on the solar tides. The effect of the 18.6-year nodal terms on the tidal range at Newlyn is shown in Figure 3.2, where the standard deviation of the sea level variations shows a clear 18.6-year cycle.

The importance of nodal terms can be further illustrated by the variation in the amplitude of $\mathbf{M_2}$ in the Equilibrium Tide. If analyses are made individually for each year throughout a nodal period of 18.6 years, the value of the amplitude will increase and decrease from the mean value by about 3.7 per cent. However, the diurnal constituents that represent the changes in lunar declination have the largest nodal variations: the $\mathbf{O_1}$ amplitude varies by 18.7 per cent; the $\mathbf{K_1}$ amplitude by 11.5 per cent; the $\mathbf{K_2}$ amplitude by 28.6 per cent.

At this stage in the development it is useful to pause and check that our ideas for a tidal analysis scheme are realistic. We can do this by checking that the biggest constituents of the Equilibrium Tide are also the most important in the observed tides, and that further development of the Equilibrium Tide will be relevant for tidal analysis. Figure 3.3 shows the relationships between the Equilibrium Tide amplitudes and phases, and those at Newlyn obtained by harmonic analysis of observations. The general trend across the semidiurnal tidal species in amplitude ratios and phase lags confirms that the relationships between Newlyn tidal harmonics and the Equilibrium Tide are generally similar for constituents of similar angular speed. We will return to this similarity in Section 3.4.

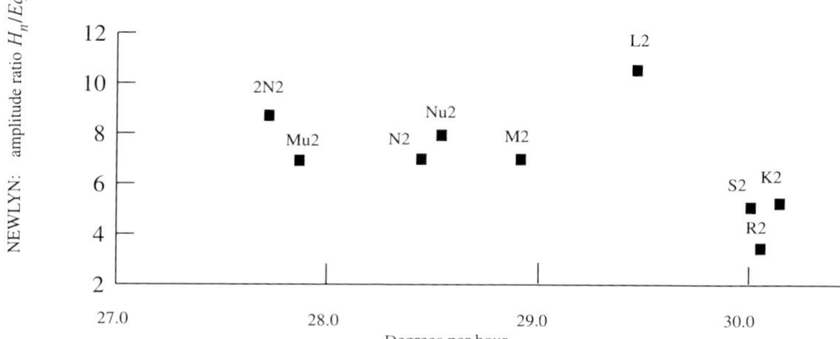

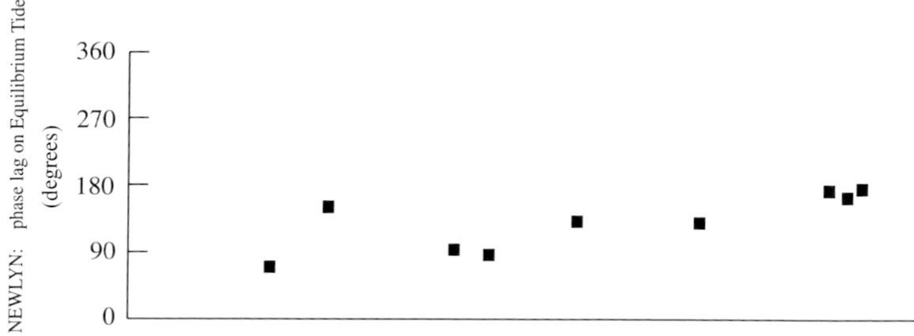

Figure 3.3. For Newlyn, the ratios of the observed constituent amplitudes (H_n) to the Equilibrium Tide amplitudes (*Eq*, for latitude 50°) and the phase lag for the principal constituents of the semidiurnal tidal species. There are systematic differences in the ocean response, but after allowing for the trends the responses are generally consistent.

3.3.2 Application of harmonic analysis

The next stage in the application of harmonic analysis is to fit our chosen harmonic constituents to sea level observations in a general form of Equation (3.2). We do this to determine the values of H_n and g_n for the gauge site. In the harmonic method of analysis we fit a tidal function to the sea level observations:

$$T(t) = Z_0 + \sum_N H_n f_n \cos[\omega_n t - g_n + (V_n + u_n)] \tag{3.5}$$

where the unknown parameters are Z_0 and the series of constituent amplitudes and phases (H_n, g_n). Z_0 is included here as a variable to be fitted in the analysis, but generally shown separately, as in Equation (3.1). The f_n and u_n are the nodal adjustments and the terms $\omega_n t$ and V_n together determine the phase angle of the Equilibrium constituent. V_n is the Equilibrium phase angle for the constituent at the arbitrary time origin. The accepted convention is to take V_n as for the Greenwich Meridian and to take t in the standard time zone of the observation station.

The fitting is adjusted so that $\Sigma S^2(t)$, the square of the residual differences between the observed $\mathbf{O}(t)$, and computed tidal levels:

$$\mathbf{S}(t) = \mathbf{O}(t) - \mathbf{T}(t)$$

has its minimum value when summed over all the observations. The least-squares fitting procedure involves matrix algebra which is outside the scope of this account (standard sub-routines are available in mathematical software packages), but schematically the equations may be written:

$$\underset{\text{known}}{[\text{observed level}]} = \underset{\text{known}}{[\text{Equilibrium Tide}]} \times \underset{\text{unknown}}{[\text{empirical constants}]} \qquad (3.6)$$

The tidal variation function T(t) includes a predetermined finite number of N harmonic constituents, depending on the length and quality of the observed data. Typically, for a year of data $N = 60$, but in shallow water more than a hundred constituents may be necessary. The choice of which constituents to include depends on the relative amplitudes in the Equilibrium expansion of the astronomical forcing (Table 3.3 and Figure 3.1), and perhaps on local knowledge. However, the Equilibrium amplitudes are not themselves part of the computations.

Least-squares fitting has several useful properties, as follows.

- Gaps in the data are permissible. The fitting is confined to the times when observations were taken. Analyses are possible where only daylight readings are available, or where the bottom or top of the tidal range is missing.
- Any length of data may be treated. Usually complete months or years are analysed.
- No assumptions are made about data outside the interval to which the fit is made.
- Transient phenomena are eliminated – only variations with a coherent phase at tidal frequencies are picked out.
- Although fitting is often applied to hourly values, analysis of observations at other time intervals is also possible.

There are certain basic rules for deciding which harmonic amplitudes and phases are to be determined in analysis. In general the longer the period of data to be included in the analysis, the greater the number of constituents that may be independently determined. One selection criterion often used is due to Rayleigh, which requires that only constituents separated by at least a complete period from their neighbouring constituents over the length of data available should be included. Thus, to determine $\mathbf{M_2}$ and $\mathbf{S_2}$ independently in an analysis requires:

$$360/(30.0 - 28.98) \text{ hours} = 14.77 \text{ days}$$

This is the spring–neap tidal period. In the previous section we discussed how tidal *groups* can normally be resolved from a month of data, whereas tidal *constituents* need a year of data for separate resolution. This minimum length of data necessary to separate a pair of constituents is called their synodic period. It can been argued that the Rayleigh criterion is unnecessarily restrictive where instrumental noise and background meteorological noise are low, but in practice, the Rayleigh criterion is a good guide for tidal analyses of continental shelf data from middle and high latitudes. Finer resolution of constituents is feasible in ideal conditions where noise from weather effects is small, such as at tropical oceanic sites.

When choosing which values of ω_n to include as tidal constituents in an analysis, scrutiny of tidal analyses from a nearby Reference Station is helpful. Where the data length is too short to separate two important constituents, it is usual to relate the Reference Station amplitude ratio and phase lag of the weaker term to the amplitude and phase of the stronger term. If a local Reference Station is not available, relationships in the Equilibrium Tide may be used instead.

It is also necessary to observe another basic rule of time-series analysis, related to the frequency at which observations are made. The Nyquist criterion states that only terms having a period longer than twice the sampling interval can be resolved. In the usual case of hourly data sampling, this shortest period is two hours, so that resolution of $\mathbf{M_{12}}$ would just be possible. In practice this is not a severe restriction except in very shallow water, where sampling more frequently than once an hour is necessary to represent the tidal curves.

3.3.3 Accuracy of tidal constituents

Some people call the tidal parameters derived from a harmonic tidal analysis *tidal constants*, although it is more correct to call them *tidal constituents*, as we do. Tidal constituents do vary slightly among different analyses of data from different times at the same site. Nevertheless, we expect these variations to be small. Reasons for the variability of the tidal constituents include analysis limitations due to non-tidal energy at tidal frequencies, inconsistencies in the measuring instruments and real oceanographic modulations of the tidal behaviour. The tidal characteristics of a site can change slowly due to both natural (silting) and artificial effects (dredging and construction). Analyses of individual months of tidal data from the same location invariably show small variations in the tidal constituent amplitudes and phases about some mean value, even allowing for nodal changes.

As an example, Figure 3.4 shows the variability of Newlyn $\mathbf{M_2}$ amplitude and phase, from analyses of individual years of data over a long period. Theoretical nodal effects have been removed in the analysis. Over 85 years the amplitudes have a standard deviation of 0.009 m

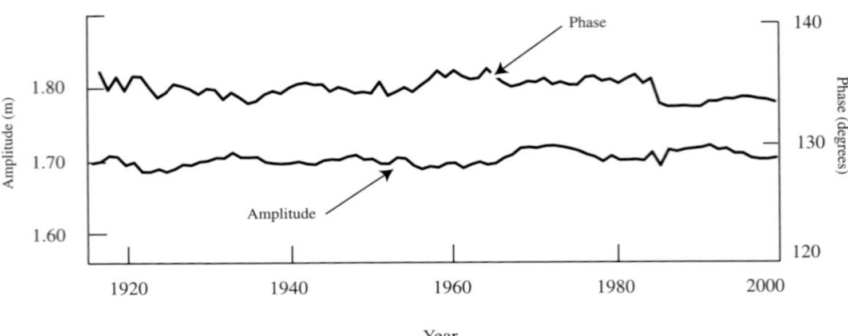

Figure 3.4. The amplitudes and phases of the principal semidiurnal lunar tide (**M_2**) at Newlyn are plotted each year over an 85-year span. This shows the uncertainty associated with harmonic constituents determined from analysis of a single year of data. See the text for further explanations of the variability.

about a mean of 1.706 m and the phases have a standard deviation of 0.8° (less than two minutes) about the mean of 134.6°. The reduction in the phases after 1984 is because a new gauge was installed outside the old stilling well; the old well had a very constricted entry hole (see Section 1.4). Close examination also shows a residual 18.6-year cycle out of phase with the Equilibrium nodal modulations; this is because in many places, including Newlyn, the full theoretical 3.7 per cent Equilibrium Tide modulation allowed for in the analyses is reduced in the real situation by shallow-water effects. There is also evidence of a small increase in the **M_2** and **M_4** amplitudes, which may be a local effect. The box on tidal changes at the end of Chapter 4 looks at trends in ocean tides and tidal constituents in further detail.

3.3.4 Harmonic equivalents of some non-harmonic terms

The harmonic constituents can be related to some of the common non-harmonic terms used to describe the tides, developed in Section 2.4 and listed in the Glossary. The most common of these describes the spring–neap modulations of the tidal range that are represented by the combination of the principal lunar and principal solar semidiurnal harmonics:

$$Z_0 + H_{M_2}\cos(2\omega_1 t - g_{M_2}) + H_{S_2}\cos(2\omega_0 t - g_{S_2})$$

where the time zero is at syzygy, when the moon earth and sun are in line at new moon or full moon.

The maximum values of the combined amplitudes are:

$$\text{mean high water springs} = Z_0 + (H_{M_2} + H_{S_2})$$
$$\text{mean low water springs} = Z_0 - (H_{M_2} + H_{S_2})$$

The minimum values are:

$$\text{mean high water neaps} = Z_0 + (H_{M_2} - H_{S_2})$$
$$\text{mean low water neaps} = Z_0 - (H_{M_2} - H_{S_2})$$

In very shallow water it is necessary to include additional shallow-water tidal constituents, such as **M_4** and **MS_4** (see Section 5.2) to represent the spring-neap cycle.

The relative importance of the diurnal and semidiurnal tidal constituents is sometimes expressed in terms of a form factor (shown in Table 3.1), derived from the harmonic constituent amplitudes:

$$F = \left(\frac{H_{K_1} + H_{O_1}}{H_{M_2} + H_{S_2}} \right) \tag{3.7}$$

In terms of the form factor, F, the tides may be roughly classified as:

$F = 0$ to 0.25	semidiurnal form
$F = 0.25$ to 1.50	mixed, mainly semidiurnal
$F = 1.50$ to 3.00	mixed, mainly diurnal
$F =$ greater than 3.0	diurnal form

The tidal predictions plotted in Figure 2.1 have the form factors as shown in Table 3.1. It is evident that ports with very different tidal regimes can have the same form factor, which limits its value as a useful description of tidal regimes.

Tides have been classified in various other general ways that can be related to the tidal constituent amplitudes. Some coastal scientists still use a very crude classification: tides with a range greater than 4 m are called *macrotidal*; those between 2 and 4 m are called *mesotidal*; and those less than 2 m are called *microtidal*. Several other non-harmonic tidal descriptors such as highest and lowest astronomical tide, and mean high and mean low water springs have been developed to describe local tidal regimes; these are defined in the Glossary.

3.3.5 Analysis of satellite altimetry data for harmonics

So far we have discussed the analysis of a series of sea level observations at a single site. Figure 3.5a shows that almost all the fixed site locations for observations are on continental coasts or ocean islands. Maps of ocean tides also need data in the open oceans, away from land. Satellite altimetry gives a much more uniform coverage of the global ocean, as shown in Figure 3.5b. To obtain local tidal harmonic constituents from these satellite tracks requires very different analytical techniques.

As shown in Section 1.4.4, careful corrections are necessary in order to obtain the heights of sea level above the reference ellipsoid from altimetry data (h_2, see Figure 1.9a). After processing the data consists

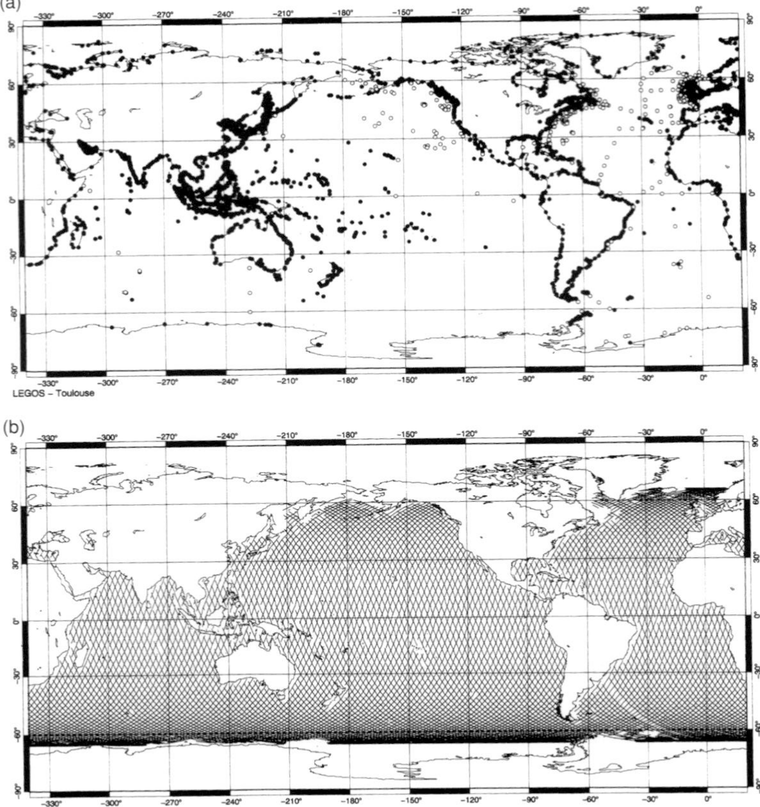

Figure 3.5. Satellite and *in situ* sea level measurements form a complementary dataset (from Fu and Cazenave (2001), with permission). (a) For the location of *in situ* observations collected since the beginning of the last century, restricted to continental coasts and islands (solid black circles) and a few open ocean sites (open circles). (b) The distribution of ground tracks of the TOPEX/Poseidon mission along which sea levels are measured every ten days.

of a series of levels (h_2) at specified positions and times. Figure 3.5b shows the ground tracks of the TOPEX/Poseidon mission, along which sea level heights are measured every ten days. Over the ten-day repeat cycle almost a million levels are recorded at one-second intervals. Of course not all of these are over the ocean.

Because the measurements of sea level made by an orbiting satellite (sometimes called *sea surface height* or SSH) at any particular location are spaced over long intervals, techniques for analysing and extracting the tidal variations are quite different from those used for long time series of sea level at a single point. Most oceanographers want sea levels for calculating currents (we will discuss this in more detail in Chapter 7). For them the tidal variations are noise to be removed from their signal. Fortunately, once ocean tides have been accurately determined, their coherence and predictability means that the tidal 'noise' can be systematically removed from future satellite altimetry records.

First it is necessary to analyse the satellite data for the tidal components. We shall see in Chapter 4 that tides in the world's oceans behave like long waves and are similar over a large area. Technically we can say that they have a high degree of spatial coherence. Many of the methods for mapping the harmonic constituents of the tides from altimetry data make use of this coherence. This is done by fitting theoretical curves over long tracks of the satellite, or by aggregating several values within a wider area. One of the more powerful techniques compares satellite observations of sea level with those generated by hydrodynamic tidal models, which are then adjusted to give better agreement. The global tidal charts in Chapter 4 were prepared in this way.

There are limitations because TOPEX/Poseidon only orbits to 66° latitude north and south of the equator, and because altimeter coverage in shallow near-shore waters, where tides change over shorter distances, is restricted. Nevertheless, the results shown in Figures 4.7 and 4.8 are in close agreement with those obtained by several alternative tidal analysis methods applied to altimeter data, and with *in situ* observations. In comparative tests at 69 island and deep-sea sites the standard deviation of the difference between predictions based on the best altimeter models and the observed sea levels was around 3.5 cm. The corresponding figure for differences between predictions based on traditional harmonic analysis of data and observations at the same sites was around 2.5 cm.

When removing ocean tides from the altimeter signal it is important to remember that tides measured by satellites include the tidal movements in the solid earth. Ground-based tide gauges measure only the ocean tide component since the gauges move up and down with the earth. Both the solid earth and ocean tides must be removed from altimeter data for studies of the dynamic sea surface height variations associated with large-scale ocean circulation.

3.4 Response analysis

There is an alternative, powerful technique for tidal analysis, called the response method, which we introduced briefly in Figure 3.3. The basic ideas involved in response analysis are common to many activities.

A system, sometimes called a 'black box', is subjected to an external stimulus or input. The output from the system depends on both the input and the system response to that input. The response of the system may be evaluated by comparing the input and the output at various forcing frequencies. Many systems can be described in these terms. Financially, one might ask how the economy responds to an increase in interest rates, and measure this response in terms of Gross Domestic Product. In engineering, the response of a bridge to various wind speeds and directions may be monitored in terms of its displacement or its vibration. Mathematical techniques for describing system responses have been developed and applied extensively in electrical and communication engineering. As an example in our own area, we have already considered the response of a stilling well to external waves of different amplitudes and periods (Section 1.4.2) in terms of the levels in the well.

In tidal analysis we have as the *input* the Equilibrium Tidal potential, which as we have seen consists of harmonics at certain well-defined frequencies, related to the movements of the moon, earth and sun. The *system* is the ocean, and we seek to describe its response to gravitational forces generated by the Equilibrium Tide. The tidal variations of sea level measured at a particular site may be considered as the *output* from the system. This 'response' treatment has the conceptual advantage of clearly separating the astronomy (the input) from the oceanography (the 'black box').

When the amplitude and phase responses at a coastal site have been computed by harmonic analysis, it is easy to calculate approximate responses from the harmonic equivalents in the Equilibrium Tidal potential, as shown in Figure 3.3. The effectiveness of the response analysis is based on the similarity of the responses at close but different frequencies, which is called the *credo of smoothness*. However, a proper response analysis is based on more formal procedures, which are beyond the scope of this account.

If we consider that a good tidal analysis represents the observations with the smallest possible number of parameters, then the response procedure is superior to the harmonic approach. Typically, the response technique can account for slightly more of the total variance than the harmonic method can accommodate, using fewer than half the number of harmonic parameters. As a research tool the response method offers other powerful opportunities, including working with shorter records, the identification of additional forcing inputs and the ability to make subtle choices of factors for inclusion. A typical response analysis will have the gravitational input, the solar radiation input and a series of shallow-water interactive inputs.

By comparison, the harmonic analysis approach offers few alternatives, apart from the addition of extra constituents, and little scope for development. The laborious adjustments we have had to introduce for nodal variations illustrate this restriction. Nevertheless, the conceptually simpler harmonic analysis is generally used for routine tidal predictions, and it is normal practice to present maps of tides in terms of the harmonic amplitudes and phase lags for individual constituents, especially for $\mathbf{M_2}$.

Many people continue to regard tidal analysis and prediction as a black art; however, tidal analysis is neither as difficult nor as mysterious as they imagine. It is important to decide which level of complexity is appropriate for a particular application. Provided that certain basic rules relating to the data length and to the number of independent parameters demanded in an analysis are followed, satisfactory results are obtained. Inexperienced analysts often go wrong when they ask for too much from too little data. However, it is true that to extract the maximum tidal information from a record, for example by exploiting the complimentary aspects of harmonic and response analysis techniques, further experience and informed judgement are necessary.

3.5 Analysis of currents

Currents are intrinsically more difficult to analyse than elevations because they are vector quantities that require more parameters for a proper description. Also, measurements of currents usually show a larger proportion of non-tidal energy than elevations, which makes the errors in the estimated tidal components bigger. A third restriction is the difficulty and expense of making current measurements offshore, which means that there is less data to analyse. Although the techniques used for analysing levels are also available for dealing with currents, less stringent standards must be applied for these reasons. Currents are traditionally measured in terms of speed and direction of flow clockwise from the north. Mathematically it is usual to separate current components into a positive north and positive east direction, V and U, and to analyse these separately for harmonic constituents. When the U and V components of each constituent are recombined, they can be plotted as the ellipse traced by the end of the current vector. In a channel (see Section 5.4) where the flow is positive and negative along a single fixed axis, the same method as for levels can be used.

3.6 Tidal prediction

3.6.1 Reference or Standard Stations

Predictions for Standard or Reference Stations are prepared directly from the astronomical arguments, using sets of local harmonic constituents previously determined by analysis. The procedure, which reverses the methods of harmonic analysis, begins with the astronomical coordinates of the moon and sun and then computes the levels for each harmonic constituent from the known amplitude and phase relationships. The left-hand side of Equation (3.6) is now the unknown factor. The full set of harmonics is added together to give the prediction of total tidal levels.

Usually, the greater the number of constituents included in this process, the more reliable are the predictions. Figure 3.6 compares the

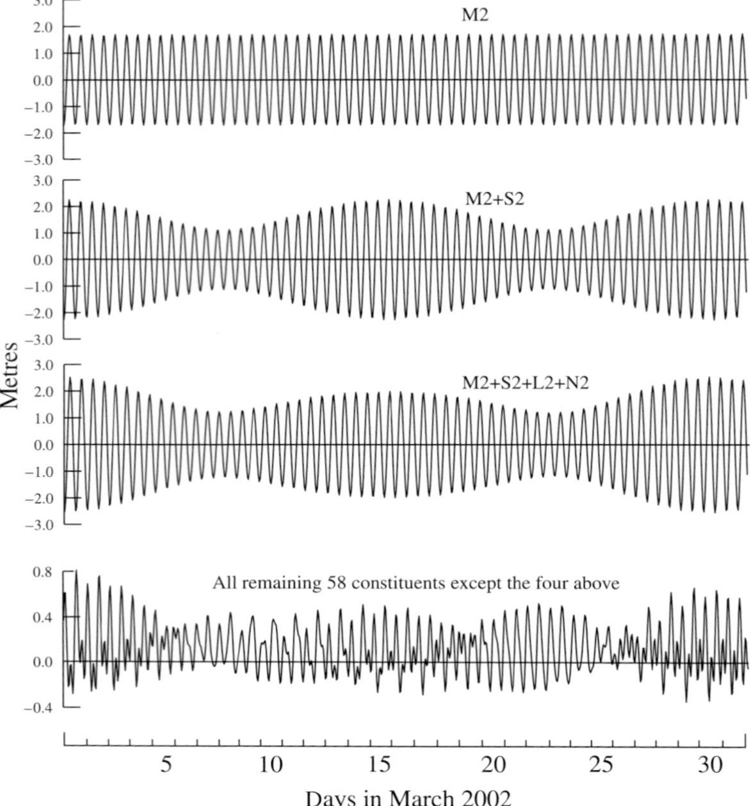

Figure 3.6. The effects of predicting tides with $\mathbf{M_2}$, $\mathbf{M_2} + \mathbf{S_2}$ and $\mathbf{M_2} + \mathbf{S_2} + \mathbf{N_2} + \mathbf{L_2}$ at Newlyn for March 2002. $\mathbf{M_2} + \mathbf{S_2}$ gives the spring–neap modulation. Adding $\mathbf{N_2} + \mathbf{L_2}$ allows for the varying distance of the moon from the earth during a lunar month. The bottom trace shows the additional contribution to a full prediction of a further 58 constituents, including diurnal and shallow-water terms.

levels predicted at Newlyn for March 2002 (see Figure 2.1) by $\mathbf{M_2}$ alone, by adding $\mathbf{S_2}$ and then by adding $\mathbf{N_2} + \mathbf{L_2}$. The spring–neap changes are introduced by $\mathbf{S_2}$ and the effects of lunar distance changes by $\mathbf{N_2} + \mathbf{L_2}$. The bottom trace shows the contribution to a full tidal prediction of a further 58 terms: diurnal and shallow-water species constituents are included. Typically, published predictions for Reference Stations are based on 60 or more constituents.

Authorities prepare the tidal predictions about two years ahead of their effective date, to allow time for printing and publication. Alternatively, predictions are now available instantaneously via the Internet. Although predictions for publication could be prepared several years ahead, this is not usually done until about two years before, because the latest datum or time zone changes need to be incorporated. Normally the tidal constituents change very little over several tens of years, as discussed in Section 3.3.3. If there are major port developments it is advisable to identify any changes that may have affected the tidal regime, by making a new analysis of a year of data.

3.6.2 Secondary or Subordinate Stations

Tidal predictions are available in many forms. Several Internet sites give predictions and sometimes plots of future tides, though some restrict availability to a few days ahead for copyright reasons. Software is available for portable computers to make predictions (See http://publishing.cambridge.org/resources/0521532 183/for updates on software and internet sites of interest.). Each year sets of Tide Tables are published covering many Reference Ports worldwide. Eventually information available on the Internet may make them obsolete.

Although these published annual Tide Tables contain daily high and low water information for a hundred or more stations, these are only a fraction of the places for which tidal information is available and for which predictions are needed. It would be impractical to prepare and publish daily predictions directly from the astronomical arguments for all of those places. Instead a simpler indirect approach is often used. Time and height differences are given, which allow adjustment to the published values at Standard Stations, to give predictions for the Secondary or Subsidiary Stations.

The tables published by the United States NOAA contain data for more than 6000 Subsidiary Stations. The British Admiralty also publishes data for several thousand Secondary Ports. Of course the term "Secondary Port" does not imply secondary importance as a *port*: it signifies only that the tidal predictions are based on another location for which predictions have been prepared directly from the astronomical information. The same port may be a Standard Port in one publication but a Secondary Port in another.

The US tables give time adjustments for high and low water, together with height adjustments and in some cases scaling factors. The British tables are more elaborate, allowing both the time and the height factors to be adjusted between spring and neap tidal conditions. Tables 3.4 and 3.5 show some computations for tides at Monterey, California, USA, and at Burncoat Head, Nova Scotia, Canada, as published by the National Ocean Service of NOAA.

The Standard Port chosen may not be the nearest to the Secondary Port – the most suitable standard will be a port which has similar tidal characteristics. For this reason the British Admiralty Tide Tables choose to refer some Japanese ports where semidiurnal tides are dominant to the Australian port of Darwin, which has similar tides, rather than to the nearer Japanese Standard Ports which have mixed tides. Secondary parameters may be based on analyses of long periods of data, but a month of observations is more usual, and normally adequate. In some remote locations the observations upon which the time and height conversions are based may have been made over only a few days.

Table 3.4. *Computations of Secondary Port times and heights at Monterey, based on San Francisco as a Standard Port and NOAA conversion factors.*

	Standard Port San Francisco		Secondary adjustment for Monterey		Computed tides at Monterey	
	Time	Height (ft)	Time	Height (ft)	Time	Height (ft)
hw	00.19 a.m.	6.2	−1.08	−0.5	11.11 p.m.	5.7 previous day
lw	06.23 a.m.	−0.3	−0.47	−0.0	5.36 a.m.	−0.3
hw	12.59 p.m.	5.4	−1.08	−0.5	11.51 a.m.	4.9
lw	6.28 p.m.	1.0	−0.47	−0.0	5.41 p.m.	1.0

Monterey tides are earlier and slightly smaller than those in San Francisco. Note that the United States publish tide heights in feet.

Table 3.5. *Computations of Secondary Port times and heights at Burncoat Head, Minas Basin, from the Canadian tables, based on Saint John as a Standard or Reference Port.*

	Standard Port Saint John		Secondary adjustment for Burncoat Head		Computed tides at Burncoat Head	
	Time	Height (m)	Time	Height (m)	Time	Height (m)
hw	0.25	8.7	+1.07	6.8	1.32	15.5
lw	6.40	−0.1	+1.11	−0.2	7.51	−0.3
hw	12.50	8.6	+1.07	6.8	13.57	15.4
lw	19.05	0.2	+1.11	−0.2	20.16	0

Burncoat Head tides are later and much bigger than at Saint John.
The secondary adjustments are those given for large tides; the Canadian tables give slightly different adjustments for medium tides.

Tides and electronic charts

New data processing and position fixing systems such as GPS are challenging the traditional ways in which ship navigators make use of sea level data. Paper charts and calculations from published Tide Tables will be around for many years to come, but there are also moves to provide real-time chart and water level information to ships entering harbour. Many argue that electronic chart systems based on vector representation, with measured sea levels automatically added to a screen display of the bottom topography of the area ahead of a ship, will make navigation in difficult waters safer. Electronic Chart Displays (ECDIS) are available to assist mariners, according to regulations approved by the International Maritime Organisation. Sea levels and currents can be included.

One small conceptual problem is that the shape and location of depth contours will change as sea levels rise and fall. However, the original contour lines and shapes can be retained if the depth that they represent is allowed to change with time. Operators of smaller boats will probably find paper charts cheaper and more convenient, and for the time being electronic charts with sea level data are seen only as an additional aid to supplement traditional navigation methods. They must prove their operational value and reliability through time and use.

Old analysis and prediction methods

An ability to predict future events, particularly those of practical importance, must inevitably have attracted some veneration to those who practised the art. The ability to predict tides could also be a source of income for those who knew the secret. The priests of the early Egyptian temples could predict the onset of the flooding of the river Nile with their nilometers. The English cleric, the Venerable Bede (AD 673–735), familiar with the tides along the coast of his native Northumbria, discussed the phases of the moon and how the tides could be calculated by the 19-year lunar cycle. By the early ninth century, tide data and diagrams, showing how neap and spring tides alternated during the month, were appearing in several tables. A later example of these is the St Albans Abbey Tide Table (ca 1250), which gives times of high water at London Bridge in terms of the age of the moon. By the seventeenth century, annual Tide Tables of times for London Bridge were produced regularly; Flamsteed, the UK Astronomer Royal, commented that his 1683 predictions were the first to give the times of two high waters each day.

In 1833, the British Admiralty published its first official annual tidal predictions (of high water times only) for four ports – Plymouth, Portsmouth, Sheerness and London Bridge. Today the harmonic method is easily applied by computers, but in earlier times the method would have been difficult to apply for routine predictions without the

Figure 3.7. A mechanical tide-predicting machine, one of the first operational analogue computers. Each pulley wheel simulates the amplitude phase and period of a harmonic constituent. These machines remained in use until around 1965 (photograph supplied by the Proudman Oceanographic Laboratory).

ingenious tide-predicting machines developed by Lord Kelvin and Edward Roberts in 1873. These machines, which were an early form of mechanical computer, consisted of a set of pulleys, one for each constituent, which rotated at different speeds scaled according to the speed of the constituent they represented (Figure 3.7). The rotations were converted into harmonic motions and summed by a means of a wire which passed over each pulley in turn. Phases and amplitudes for each constituent were set by adjusting the pulley settings for each port before running the machine. To compute high and low water times and heights for a single port for one year took about two days of work. These machines were the basis for tidal predictions until around 1965 when they were replaced by electronic computers. A year of tidal predictions can now be calculated in a fraction of a second.

Further reading

Readers interested in the development of harmonic expansions of the tidal forces should consult the original works by Doodson (1921), Doodson and Warburg (1941), Cartwright and Edden (1973) and Schureman (1976). The response analysis method is rigorously described by Munk and Cartwright (1966). Godin (1972) looks at tidal analysis in a more general context. Most of these texts assume a good mathematical background.

The methods for predicting tides, especially for Secondary or Subordinate Stations, are available on several websites, and are outlined in the

Tide Tables produced annually by Hydrographic Authorities. These publications also tabulate the time and height differences relative to standard or reference ports.

Questions

3.1 Why is it scientifically incorrect to describe a sea level measuring instrument as a tide gauge?

3.2 What is the best method of tidal analysis?

3.3 Why are measurements over a lunar month (709 hours) or a year (8766 hours) recommended for tidal analyses?

3.4 Why do the times of minimum sea level standard deviation at Newlyn in Figure 3.2 coincide with years of maximum lunar declination (1969, 1987) given in section 2.4.3?

3.5 Why are form factors of only limited value in describing the tides at a particular location?

3.6 From Figure 3.6, what would be the maximum errors if predictions were based only on the four semidiurnal constituents shown? What characteristic is seen in the errors at spring tides?

3.7 Calculate the speeds of $\mathbf{O_1}$, $\mathbf{S_2}$ and $\mathbf{K_2}$ from Tables 3.2, 3.3 and Equation (3.3).

3.8 We can allocate sea level changes to categories as in Equation (3.1). For a series of annual tidal analyses, where would we find:

- long-term changes in air pressure;
- the 18.6-year nodal effects on mean sea level;
- 18.6-year modulations in the amplitude of $\mathbf{M_2}$?

3.9 Calculate high and low water times and heights at Berkeley for the afternoon of 30 March 2002. High waters are 21 minutes later and 0.1 feet higher than at the San Francisco Reference Station. Low waters are 38 minutes later and have the same height.

Chapter 4
Tidal dynamics

The gravitational tidal forces of the moon and sun regularly raise and lower the surface of the oceans. These disturbances propagate freely in the form of waves across oceans, until they cross or are reflected at continental shelf boundaries. Eventually they dissipate through frictional energy losses in shallow water. Dynamically there are two essentially different types of tidal regime: in the wide and relatively deep ocean basins, the observed tides are generated directly by the astronomical gravitational forces; in the shelf seas that surround the oceans, the tides are driven by the spreading oceanic tidal waves.

We begin this chapter by explaining why the observed tides are very different from the theoretical Equilibrium Tide, as is evident in Figure 2.1. We will then examine and explain tidal dynamics in two stages. Firstly, we develop some useful tools for describing the behaviour of tidal waves on a non-rotating sea for simplicity, and then extend this to the real case of a sea on the rotating earth. Secondly, we apply these tools to describe the tidal patterns and behaviour actually observed in the world's oceans and adjacent shallow seas. Finally, we include a brief discussion about tidal movements inside the ocean, about which much more is becoming known. We also outline the way in which, when tidal energy moves from the deep ocean on to the shelf, it is eventually lost because of bottom friction drag which opposes tidal currents.

4.1 Tides in the real world

We have already established the main characteristics of the Equilibrium Tide. It consists of two symmetrical tidal bulges, directly opposite the moon or sun. The semidiurnal tidal ranges reach their maximum value of about 0.5 m at equatorial latitudes. The individual high water bulges track around the earth, moving from east to west in steady progression as the earth rotates. Figure 2.1 clearly shows that these theoretical Equilibrium Tide characteristics are *not* the same as those seen in the real ocean.

The observed tides in the main ocean have much larger mean ranges than the Equilibrium Tide. These are typically about 1 m, but there are considerable local differences. In general the semidiurnal tidal variations are much more important than the diurnal variations. Times of tidal high water vary in a geographical pattern that bears no relationship to the simple ideas of a double bulge. In several places the tidal variations of sea level sometimes disappear altogether.

As the tides propagate and spread from the oceans on to the shallower surrounding continental shelves, ranges larger than those of the Equilibrium Tide are observed. In some shelf seas the spring tidal range may exceed 10 m: the Bay of Fundy, the Bristol Channel and the Argentine Shelf are well-known examples of large tidal ranges. The reasons for these complicated ocean responses to tidal forcing may be summarised as follows.

- Movements of water on the surface of the earth must obey the physical laws represented by the hydrodynamic equations of continuity and momentum balance; this means that they must propagate as long waves. Any propagation of a wave, east to west around the earth in the form of an Equilibrium Tidal bulge is impeded by the north–south continental boundaries.
- In the Equilibrium Tide the high water would move from east to west at the speed of the rotating earth. Long waves travel at a speed that is related to the water depth; oceans are too shallow for this speed to keep up with the tracking of the moon or sun.
- The various ocean basins have their individual natural modes of oscillation which influence their response to the tide-generating forces. There are many resonant frequencies in the oceans. However, the oceans are nearer to resonance at semidiurnal tidal frequencies, as the observed semidiurnal tides are generally much bigger than the diurnal tides.
- Water movements are affected by the rotation of the earth. The tendency for water movement to maintain a uniform direction in absolute space means that it performs a curved path in the rotating frame of reference within which our observations are made on earth.
- The solid earth responds elastically to the imposed gravitational tidal forces and to the ocean tidal loading. As a further complication the redistribution of water mass during the tidal cycle itself affects the tidal gravitational field.

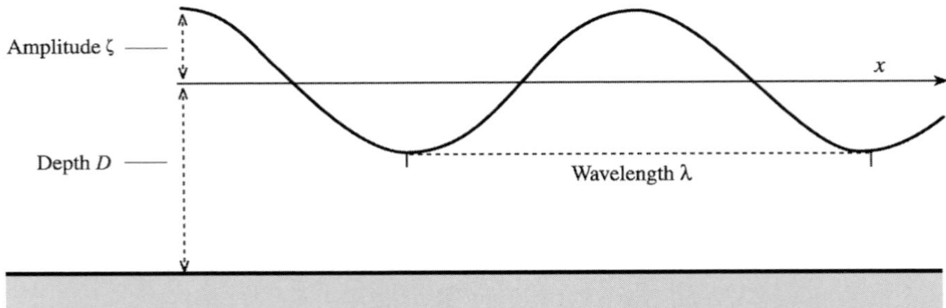

Figure 4.1. Characteristics of the one-dimensional progressive wave. The wave period is the time it takes to travel one wavelength.

4.2 Characteristics of long waves

Tides can be described in terms of the behaviour of long waves on a rotating earth. These have some simple and useful properties.

4.2.1 Long-wave propagation

Consider a wave travelling as shown in Figure 4.1. Provided that wave amplitudes are small compared with the depth and that the depth is small compared with the wavelength ($\zeta \ll D \ll \lambda$ in Figure 4.1), then the wave moves at a speed:

$$c = (gD)^{\frac{1}{2}} \tag{4.1}$$

where g is the gravitational acceleration and D is the water depth. The currents in the wave, u, are related to the instantaneous level ζ by:

$$u = \zeta \left(\frac{g}{D}\right)^{\frac{1}{2}} \tag{4.2}$$

From Equation (4.1) we can see that long waves have the special property that the speed c is independent of the frequency or wave period. The speed depends only on the value of g and the local water depth so that all waves travel at the same speed through any local area of ocean. This means that diurnal and semidiurnal tidal waves travel at the same speed. As a result, any disturbance which consists of a number of separate harmonic waves will not change its shape as it propagates – this is called non-dispersive propagation.

This non-dispersive behaviour is quite different from the behaviour of wind waves which have much shorter periods and wavelengths; they undergo dispersive propagation, with swell from distant storms travelling

Table 4.1. *Speed and wavelengths of tidal waves in water of different depths.*

	Speed		Wavelength (km)	
Depth (m)	(m s^{-1})	(km h^{-1})	Diurnal	Semidiurnal
4000	198	715	17720	8860
200	44	160	3960	1980
50	22	80	1980	990
20	14	50	1250	626

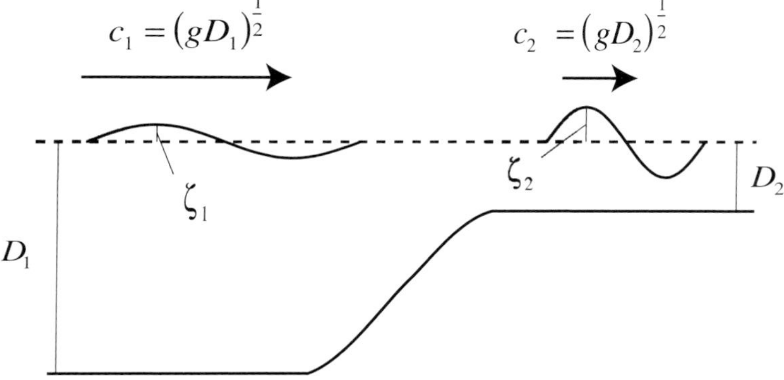

Figure 4.2. Changes in a tidal wave as it progresses from the deep ocean on to the shallower continental shelf. The wave slows down, the wavelength is shorter and the amplitude is increased.

more rapidly ahead of the shorter period waves. Table 4.1 lists some characteristic speeds and wavelengths of progressive tidal waves, at both diurnal and semidiurnal periods, which shows that wavelengths are much greater than the depths. Waves at tidal periods are long waves, even in the deep ocean, and so their propagation is non-dispersive.

In the real ocean, tides cannot propagate endlessly as progressive waves. Instead they undergo reflection at sudden changes of depth and at coastal boundaries. Figure 4.2 shows what happens to a progressive wave as it travels across the shelf edge from the deep ocean. Part of the wave energy is reflected back at the boundary and part of the wave travels on to the shelf. On the shelf the wave speed is reduced from around 200 m s^{-1} to around 40 m s^{-1}, because the depth is much less. The wavelength is also reduced because the wave period remains the same. The wave amplitude increases as the water depth reduces because the wave energy is now concentrated in a smaller area. Eventually, as the wave progresses into shallower coastal waters, the friction of the sea bed – which opposes the stronger tidal currents– reduces the tidal amplitude.

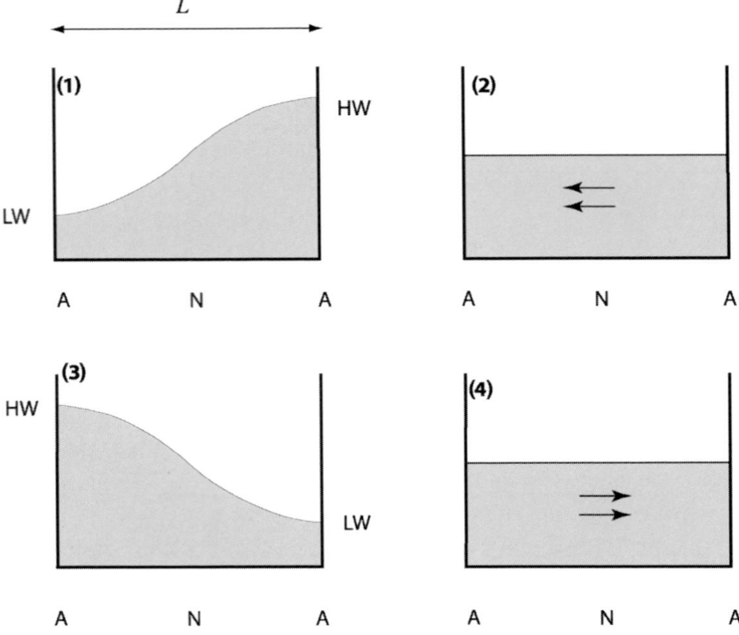

Figure 4.3. A simple standing wave oscillation in a closed rectangular box, showing nodes (N) and anti-nodes (A), for levels and currents. The sequence moves from (1) to (4) and then back to (1), in quarter-period steps.

4.2.2 Standing waves and resonance

Two equal progressive waves travelling in opposite directions result in a fixed or frozen wave motion, called a *standing wave*. This is illustrated in Figure 4.3 for a wave in a closed basin, which is one of the most effective ways of describing the behaviour of standing waves.

Consider the movement of water in a rectangular box of length L. This box may represent a domestic bath, a swimming pool, or a long narrow lake or sea. The water movements can be compared to those of a pendulum, with the least horizontal motion at high and low water ((1) and (3)) and the maximum horizontal motion at the middle of the oscillation ((2) and (4)). There is a continuous transfer from maximum potential energy (1) through a position of zero potential energy but maximum kinetic energy (2) to a second position of maximum potential energy (3). The sequence continues through a second condition of maximum kinetic energy (4) as it returns to its initial state of maximum potential energy (1).

Mathematically we can represent our standing wave as two equal progressive waves travelling in opposite directions with perfect reflections at both ends of the basin. The natural period of oscillation of the water is the time taken for a wave to leave one boundary and to return

Table 4.2. *Examples of the natural period of oscillation of water bodies in the first gravitational mode (as shown in Figure 4.3).*

Basin type	Length	Depth (m)	Period
Bath	1.5 m	0.2	2.1 s
Swimming pool	10 m	2.0	4.5 s
Loch Ness, Scotland	38 km	130	35 mins

after reflection at the second boundary. This time is:

$$\frac{2 \times \text{box length}}{(g \times \text{water depth})^{\frac{1}{2}}} = \frac{2L}{(gD)^{\frac{1}{2}}} \tag{4.3}$$

This is called the fundamental *mode* of oscillation. For the fundamental mode there is one central location of zero water level changes, called a *node*. Some examples of typical periods are given in Table 4.2.

Standing waves may also occur in a box that is closed at one end but driven by oscillatory in-and-out currents at the other open end (Figure 4.4). The simplest case is a box whose length is a quarter wavelength of the oscillation, so that the open end is at the first node (case (2) in Figure 4.4), equivalent to half the box in Figure 4.3. The current at the entrance produces large changes of level at the head of the box or basin. The natural period for this type of forced oscillation is:

$$\frac{4 \times \text{box length}}{(g \times \text{water depth})^{\frac{1}{2}}} = \frac{4L}{(gD)^{\frac{1}{2}}} \tag{4.4}$$

Although this model of an open box approximates to the tidal behaviour of many shelf sea basins, an exact quarter-wave dimension would be very unlikely. In reality, although the open boundary may lie within the node (3) or outside the node (1) as shown in Figure 4.4, the probability of tidal amplification still exists. However, if the length of the basin is only a small fraction of the tidal wavelength (4), then the amplification will be small. The semidiurnal tide in Long Island Sound, on the east coast of the United States, is an example of this type of resonance.

Systems that are forced by oscillations close to their natural period (2) have large amplitude responses (Table 4.3). The responses of oceans and many seas are close to semidiurnal resonance. In nature the forced resonant oscillations cannot grow indefinitely because of energy losses due to friction.

Also, because of energy losses, tidal waves are not perfectly reflected at the head of a basin, which means that the reflected wave is smaller than the ingoing wave. This is equivalent to a progressive wave superimposed

Table 4.3. *Lengths and depths of basins which would have quarter-wave resonance if driven by a semidiurnal* $\mathbf{M_2}$ *tide.*

Water depth (m)	Basin length (km)
4000	2200
1000	1100
200	500
100	350
50	250

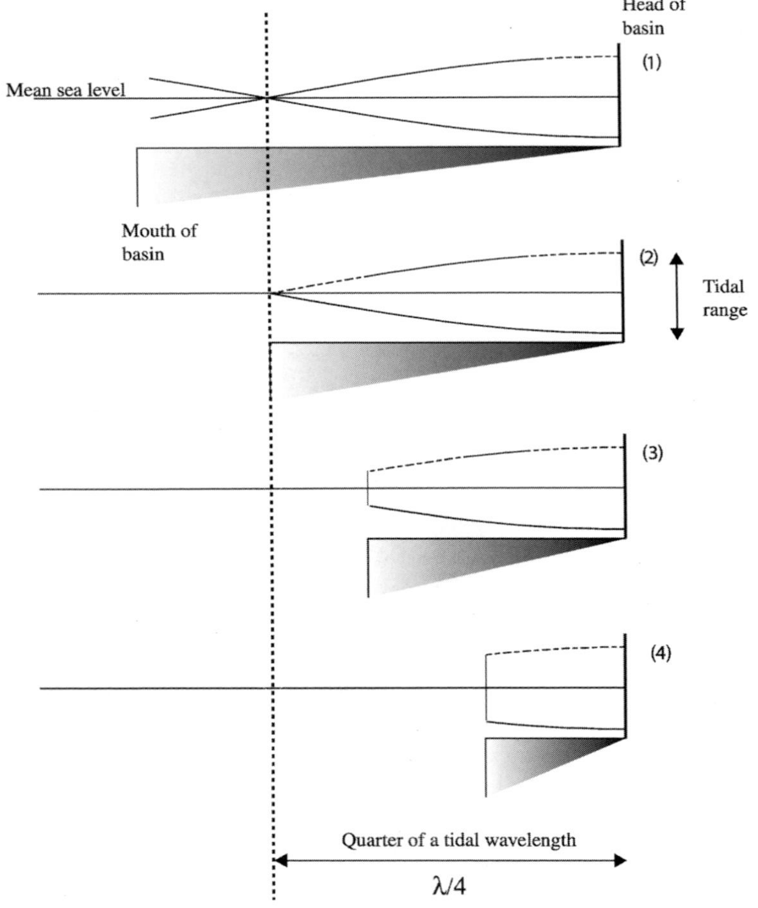

Figure 4.4. Responses of four sea basins of decreasing lengths, driven by tidal level changes at the open end, near to quarter-wave resonance. There is maximum amplification (measured as the ratio between the range at the mouth and at the head of the basin) when the basin length corresponds to a quarter of a tidal wavelength (case (2)).

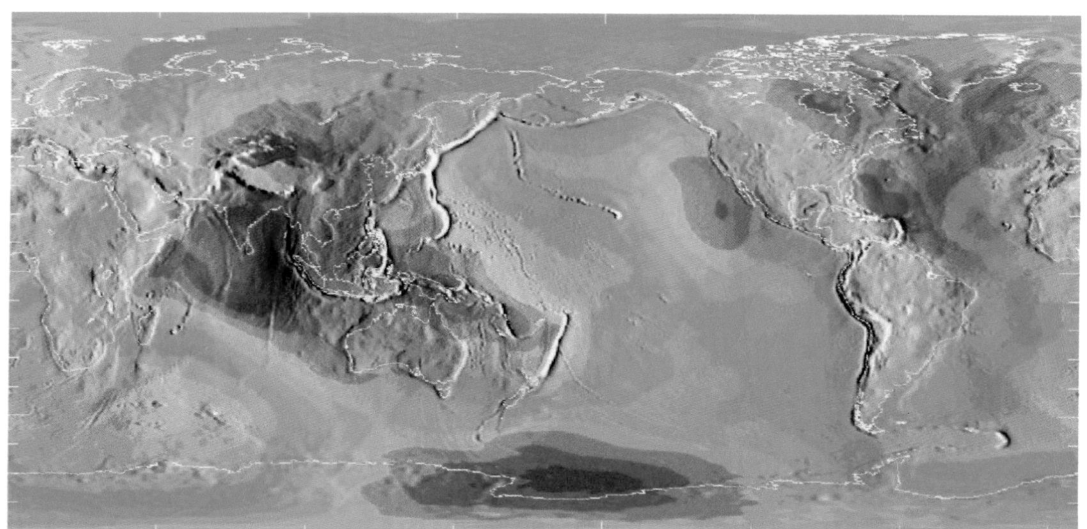

Figure 2.7. Geoid heights relative to the ellipsoid of revolution. These differences are due to uneven mass distribution in the earth. Minimum levels (−106 m, in purple) are found in the Indian Ocean. Maximum geoid levels (in orange) are about 85 m. The ocean trenches are clearly visible as a mass deficiency. This GEM 96 map was produced by the United States National Mapping Agency (NIMA) and NASA.

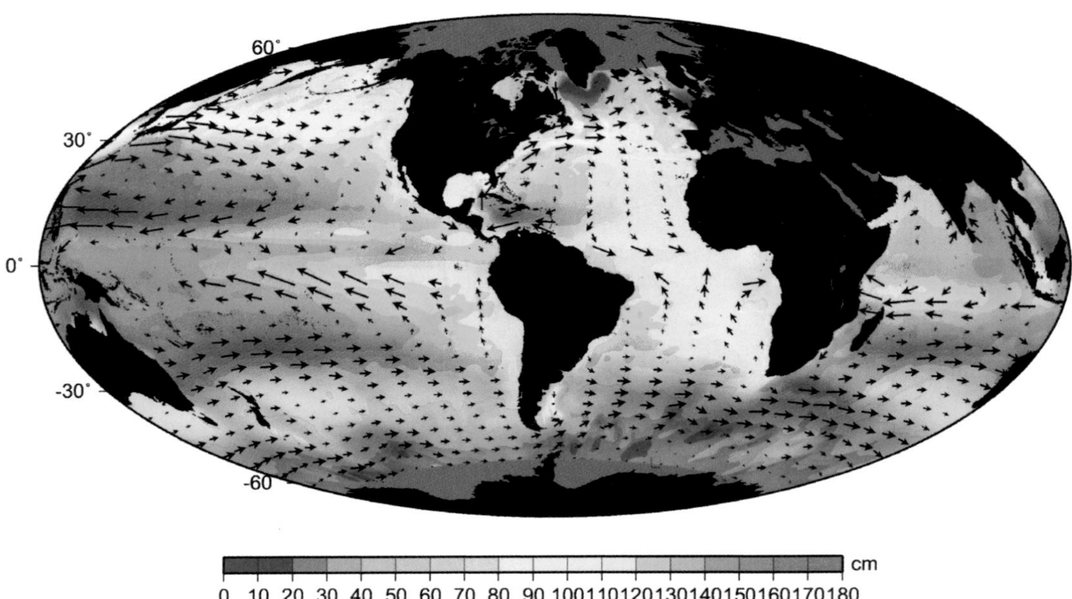

Figure 7.4. This map shows how the MSL varies from the geoid because of ocean density, currents and meteorological effects. The greatest differences are found across ocean currents (black arrows). Zero is the lowest level (courtesy Aviso-CLS).

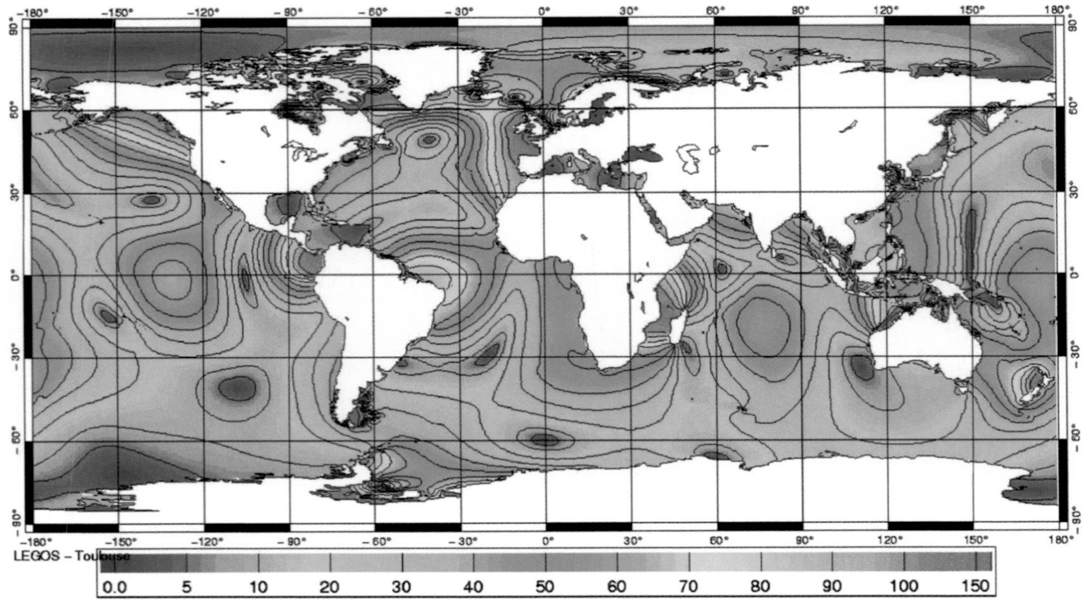

M_2 amplitude (cm)

Figure 4.7. (a) The amplitudes – note the large ranges around the amphidrome in the North Atlantic, which is close to half-wave resonance.

(b)

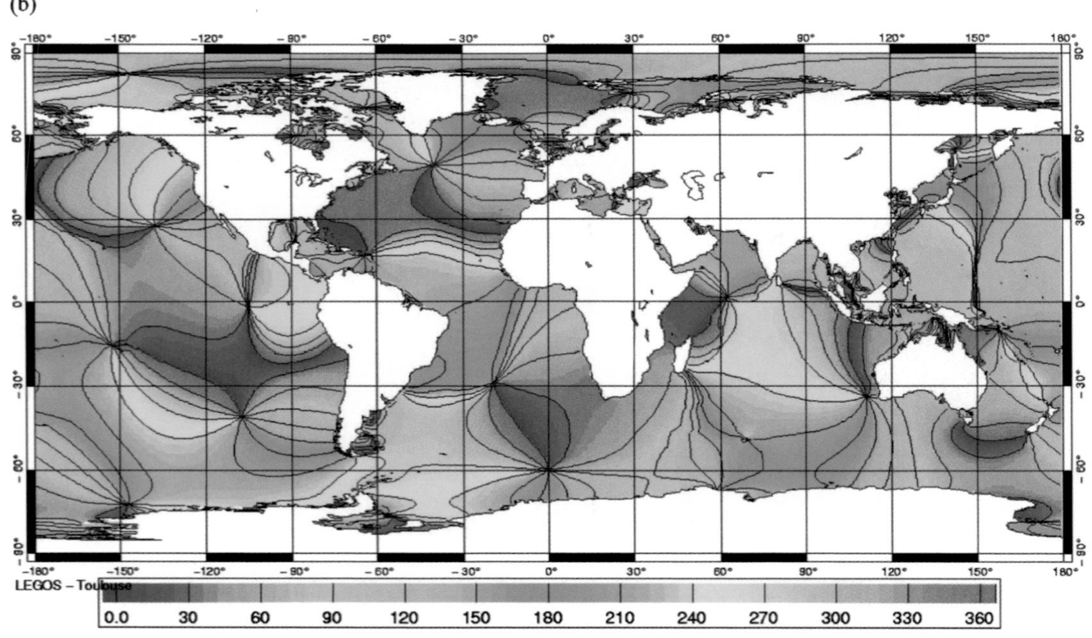

M_2 phase, g (degrees)

Figure 4.7. (b) Phases – 30° is one-twelfth of a cycle, or 62 minutes. In the northern hemisphere the tidal waves propagate in an anti-clockwise sense around amphidromic points; in the southern hemisphere there is normally clockwise propagation.

(a)

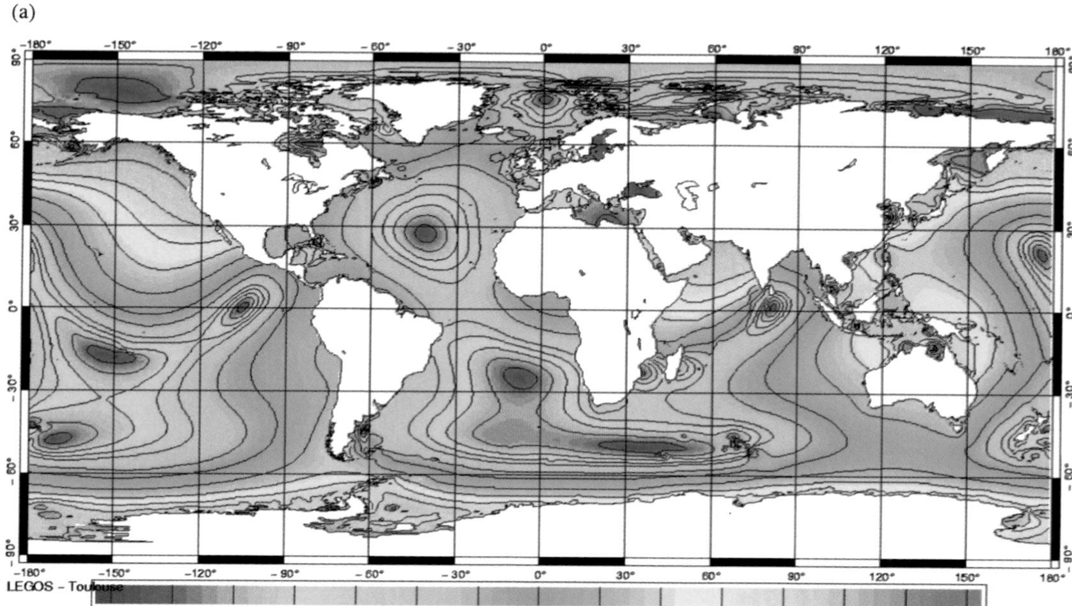

K_1 amplitude (cm)

Figure 4.8. (a) Amplitudes, which are generally less than those for semidiurnal tides.

(b)

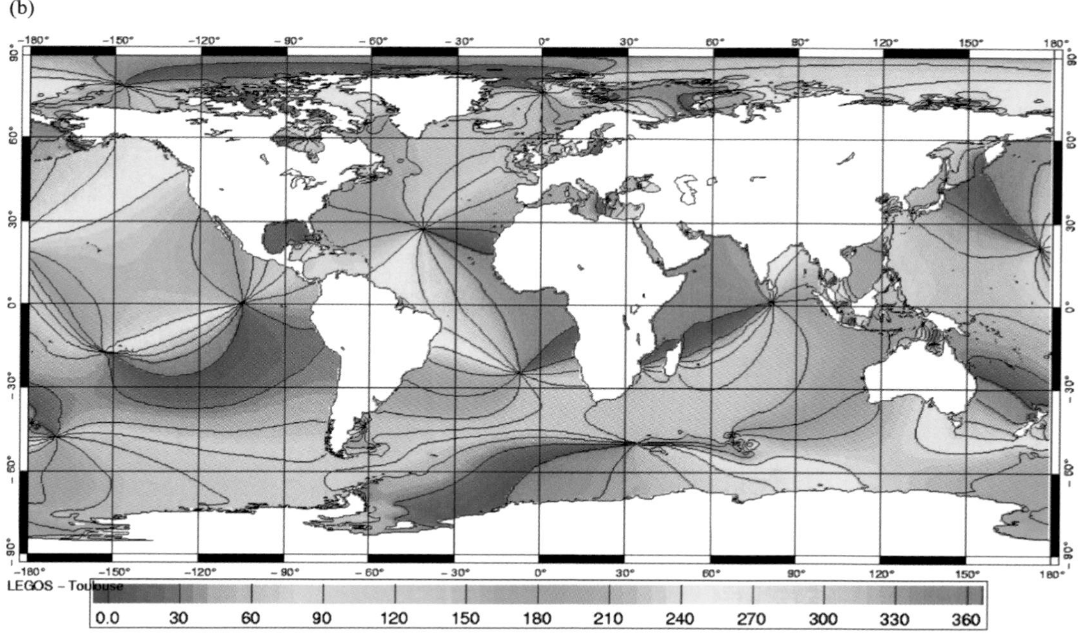

K_1 phase, g (degrees)

Figure 4.8. (b) Phases, with 30° equivalent to two hours. Because the periods and wavelengths are roughly twice those of semidiurnal tides there are fewer amphidromes. Note the Kelvin wave progression from east to west around the coast of Antarctica.

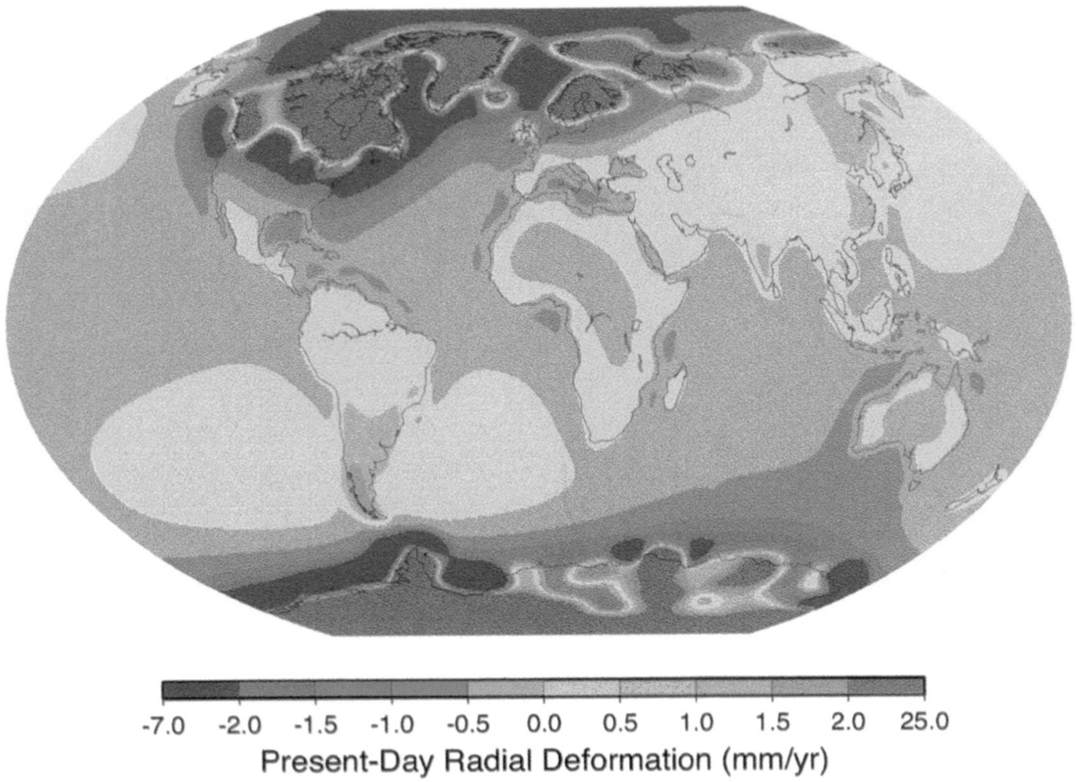

Figure 7.9. This map shows how the land is moving vertically through post-glacial adjustment after the removal of the ice sheets. It shows land is rising in the polar regions and generally falling in the tropics. The maximum uplift rates (red) reach 20–30 mm yr^{-1} in places (for example, northeast Canada). The maximum subsidence rates reach 6–7 mm yr^{-1} (for example, between Greenland and northeast Canada) (provided by Glenn Milne, University of Durham; see Mitrovica and Milne (2002)).

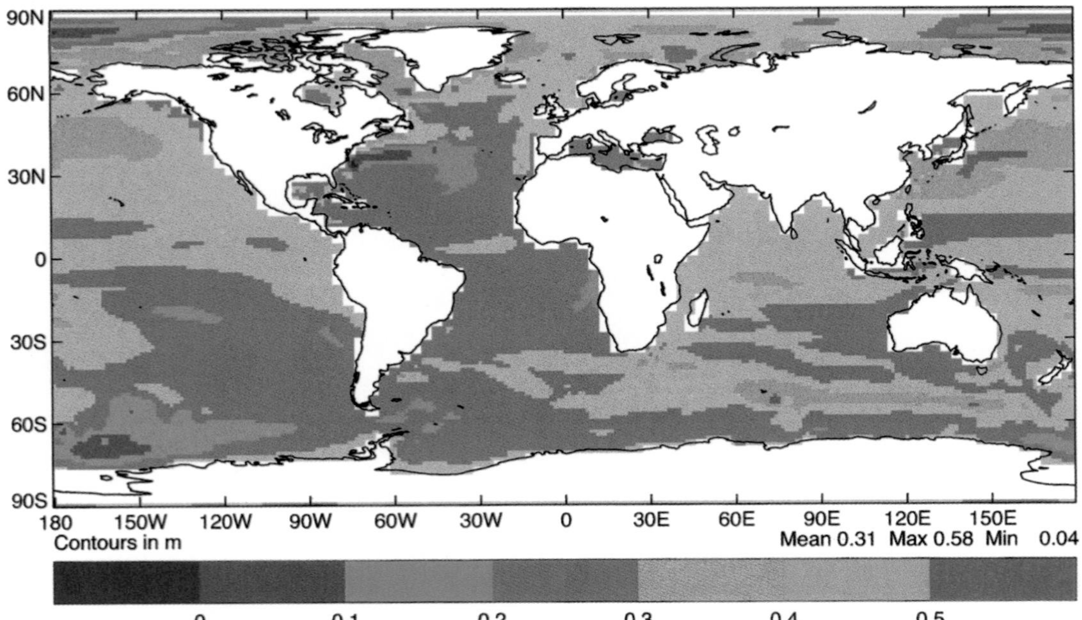

Figure 8.6. Computed sea level changes in metres between 1990 and 2090 for particular emission scenario. This map is based on the results from a coupled global–ocean atmosphere model and supplied by the UK Meteorological Office Hadley Centre. There is not yet strong agreement among the various models, except to show lower than average MSL rise in the Southern Ocean.

on a standing wave, with the progressive wave carrying energy to the head of the basin. Standing waves cannot transmit energy because they consist of two progressive waves of equal amplitude travelling in opposite directions, whose energy transmissions cancel.

4.2.3 Long waves on a rotating earth

A long progressive wave travelling in a channel on a rotating earth behaves differently from a wave travelling along a non-rotating channel. The geostrophic forces that describe the motion in a rotating system cause a deflection of the currents towards the right of the direction of motion in the northern hemisphere (and to the left in the southern hemisphere). The build-up of water on the right of the channel results in a greater wave amplitude nearer the coast; this in turn gives rise to a pressure gradient across the channel, which develops until it balances the geostrophic force, as shown qualitatively in Figure 4.5a and defined more exactly in Figure 4.5b.

The form of this wave, called a Kelvin wave, is very important for tidal studies. Although we do not need the details for this discussion, mathematically a Kelvin wave is expressed in terms of sea levels and currents by the expressions:

$$\zeta(y) = H_o \exp\left(-\frac{fy}{c}\right); \quad u(y) = \left(\frac{g}{D}\right)^{\frac{1}{2}} \zeta(y) \tag{4.5}$$

where $\zeta(y)$ is the amplitude at a distance y from the right-hand boundary (in the northern hemisphere); f is the Coriolis parameter ($f = 2\omega_s \sin\phi = 1.459 \times 10^{-4} \sin\phi$, where ϕ is the latitude), which represents the rate of rotation of the earth relative to a fixed celestial point (note that $\omega_s = \omega_0 + \omega_3 = \omega_1 + \omega_2$); c is the wave speed $(gD)^{\frac{1}{2}}$; and H_o is the amplitude of the wave at the boundary.

The effect of the rotation appears only in the factor $\exp(-fy/c)$, which gives a decay of wave amplitude away from the boundary with a length scale of $R = c/f = [(gD)^{\frac{1}{2}}/f]$, which depends on the latitude and the water depth D. This scale, called the Rossby radius of deformation, is shown in Table 4.4 for some typical depths and latitudes. At a distance $R = c/f$ from the boundary the amplitude has fallen to $0.37H_o$. At 50°N in water of 4000 m depth the Rossby radius is 1770 km, but in water only 50 m deep it is reduced to 200 km. The maximum forward currents in the direction of wave propagation coincide with high water, as for the progressive waves discussed earlier on a non-rotating sea.

Kelvin waves are not the only possible form for tidal waves on a rotating earth: a special solution gives inertial currents. Nevertheless, in practice Kelvin waves are the most important for describing the

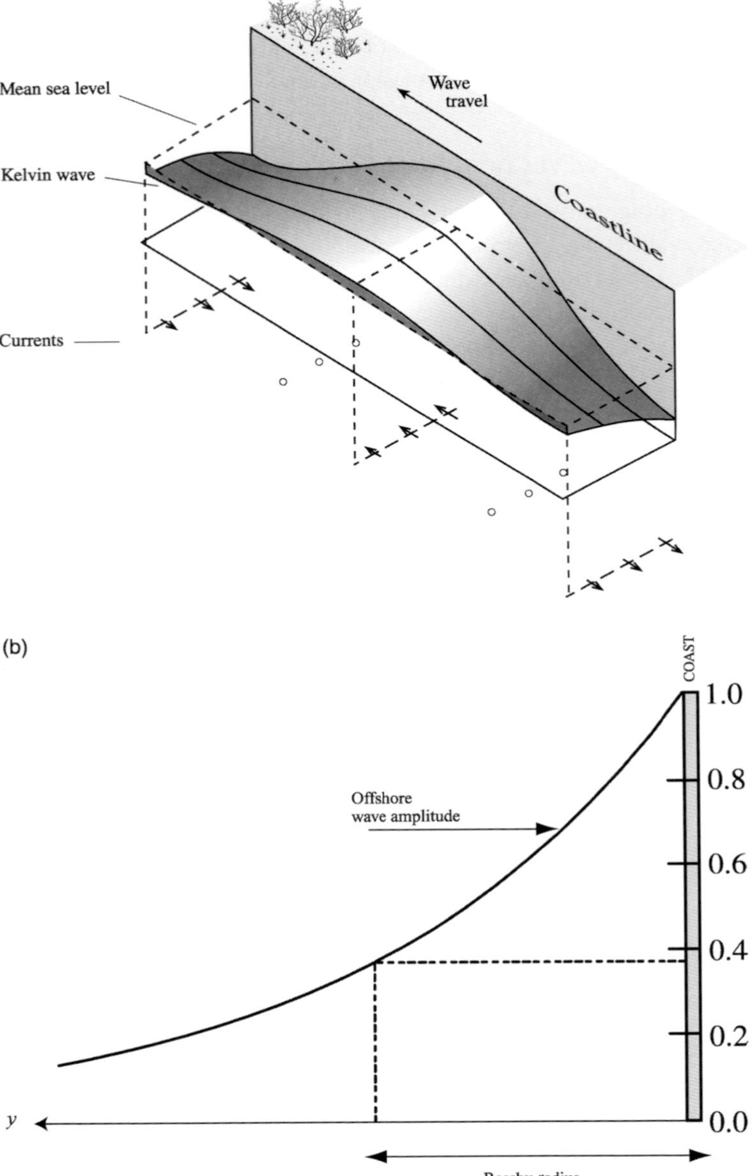

Figure 4.5. Kelvin wave dynamics. (a) A three-dimensional illustration of the elevations and currents for a Kelvin wave running parallel to a coast on its right-hand side (northern hemisphere). The normal dynamics of wave propagation are altered by the effects of the earth's rotation which act at right angles to the direction of wave travel. (b) A two-dimensional section across the Kelvin wave at high tide, showing how the tidal amplitude falls away exponentially from the coast; at one Rossby radius the amplitude has fallen to 0.37 of the coastal amplitude. The wave is travelling into the paper.

behaviour of tidal waves, and other forms are generally of only academic interest.

The behaviour of a standing wave oscillation on a rotating earth is of special interest in tidal studies. Before we consider this case it is necessary to introduce two tools that are used in mapping tidal times and

Table 4.4. *The Rossby radius (in kilometres) for different water depths and latitudes. The amplitude of a Kelvin wave is reduced by a factor of 0.37 at a distance of one Rossby deformation radius from the boundary.*

Latitude (degrees)	10	30	50	70	90
$f \times 10^{-5}$	2.53	7.29	11.2	13.7	14.6
Water depth (m)					
4000	7820	2720	1770	1450	1360
200	1750	610	395	325	305
50	875	305	200	160	150
20	550	190	125	102	96

tidal amplitudes. On charts of oceans and seas it is possible to draw lines connecting places where the times of tidal high water (or low water) are the same; we call these co-tidal lines. It is also possible to draw lines connecting places where the tidal amplitudes or ranges are the same; these are called co-amplitude or co-range lines. For a harmonic tidal constituent these are the lines joining places with the same H_n and g_n defined in Equation (3.2). As an example, for a Kelvin wave the co-range lines are parallel to the coast because the amplitude falls off exponentially; the co-tidal lines are at right angles as the wave travels along the coast.

For a standing wave in a basin on a rotating earth the co-tidal and co-range lines have a very distinctive pattern. Away from the reflecting boundary, tidal waves can be represented by two Kelvin waves travelling in opposite directions. Instead of oscillating about a nodal line, the wave can now be seen to rotate about a nodal point, which is called an *amphidrome*, as shown in Figure 4.6a. The co-tidal lines, joining times of simultaneous high water, all radiate outwards from the amphidrome. The co-amplitude lines form a set of nearly concentric circles around the centre at the amphidrome. At the amphidrome the amplitude is zero. There is no tidal phase at an amphidrome; asking for the tidal phase at an amphidrome is a little like asking for solar time at the north and south poles! Note that in Figure 4.6 the amplitude is greatest around the boundaries of the basin.

If the reflected wave is weakened, the amphidrome is displaced as in Figure 4.6b. In many cases the reflected Kelvin wave is too weak for cancellation within the ocean basin. Instead, there is convergence to a virtual inland amphidrome towards which the co-tidal lines point. These are called degenerate amphidromes; we will discuss some examples later

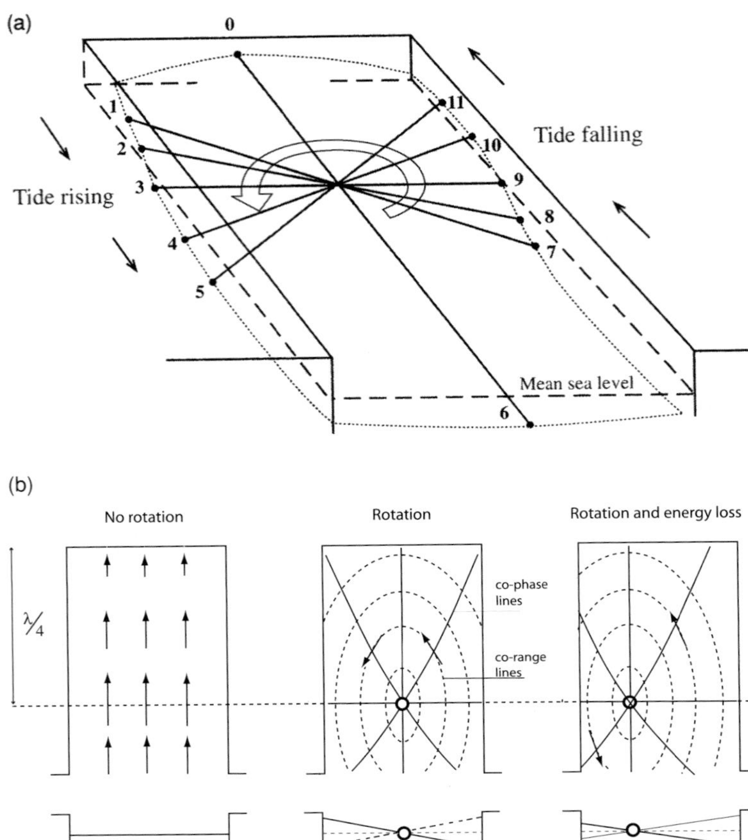

Figure 4.6. Amphidrome dynamics. (a) A three-dimensional drawing exaggerated to illustrate how a tidal wave progresses around an amphidrome in a basin in the northern hemisphere. (b) The effects of the earth's rotation on a standing wave in a basin that is slightly longer than a quarter-wavelength. With no rotation there is a line of zero tidal amplitude. Because of the earth's rotation, the tidal wave rotates around a point of zero amplitude, called an amphidromic point. In the third case, because the reflected wave has lost energy through tidal friction, the amphidrome is displaced from the centre line.

in the chapter. We can now use these ideas and tools to describe the tides actually observed in some of the oceans and shelf seas.

4.3 Ocean tides

Figure 4.7 shows a global chart of the phases and amplitudes of the principal lunar semidiurnal tidal constituent $\mathbf{M_2}$. Figure 4.8 shows the same information for the diurnal $\mathbf{K_1}$ tide. Charts of the tidal constituent waves can be produced from altimeter data, or by computer simulation from the tidal forces with adjustments to fit where there are observations. The most obvious feature of the chart is the large number of amphidromes. As a general rule these ampidromes conform to the expected behaviour for Kelvin wave propagation, with anti-clockwise rotation in the northern hemisphere and clockwise rotation in the southern hemisphere. There are exceptions, for example, the $\mathbf{M_2}$ system west of Africa in the South

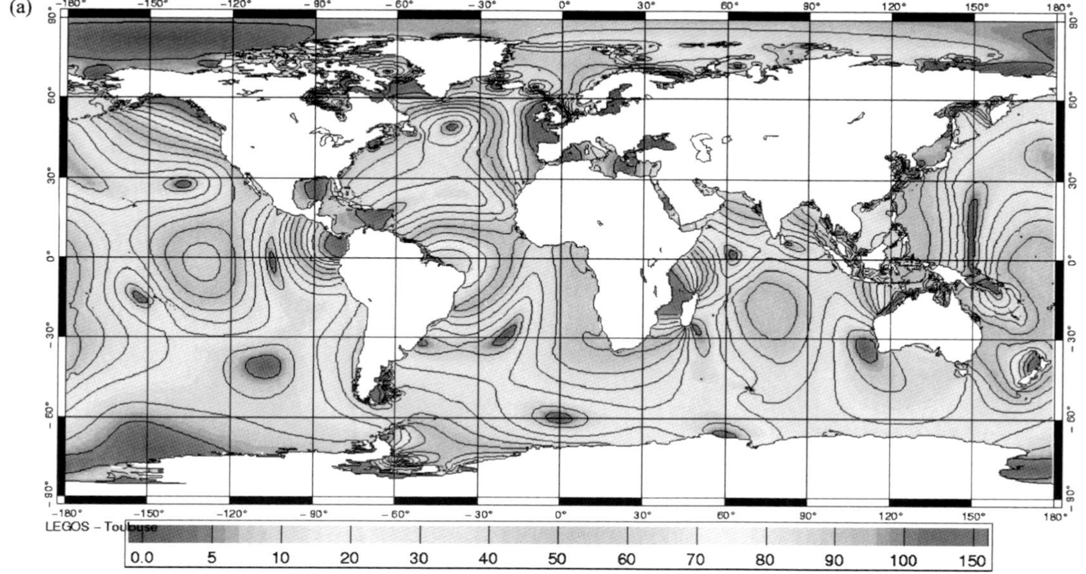

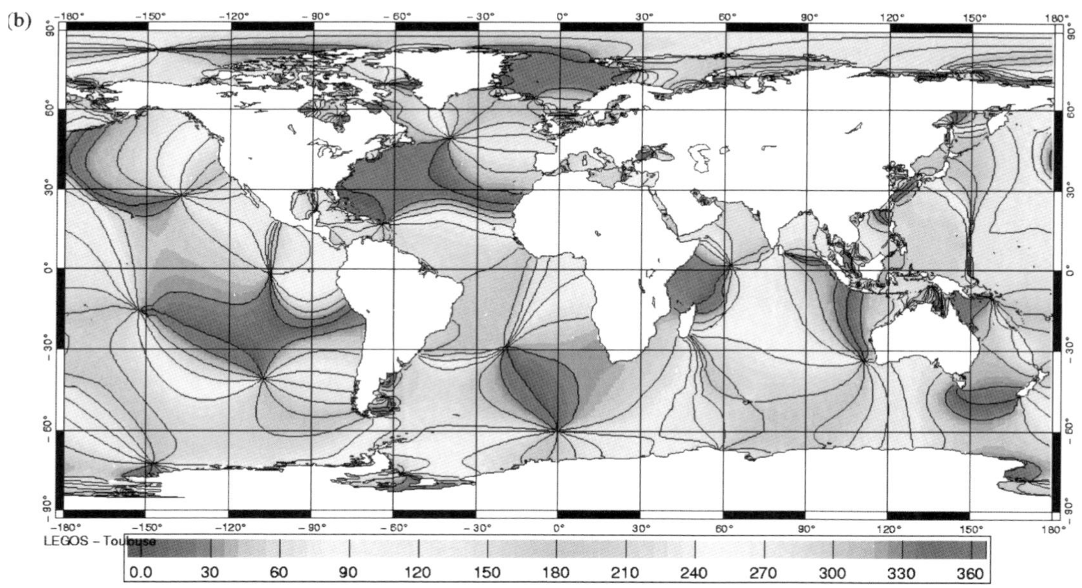

Figure 4.7. A computer-generated map of the $\mathbf{M_2}$ semidiurnal tidal wave in the global ocean. (a) The amplitudes – note the large ranges around the amphidrome in the North Atlantic, which is close to half-wave resonance. (b) Phases – 30° is one-twelfth of a cycle, or 62 minutes. In the northern hemisphere the tidal waves propagate in an anti-clockwise sense around amphidromic points; in the southern hemisphere there is normally clockwise propagation. (Supplied by Christian Le Provost.) See colour plate section.

(a)

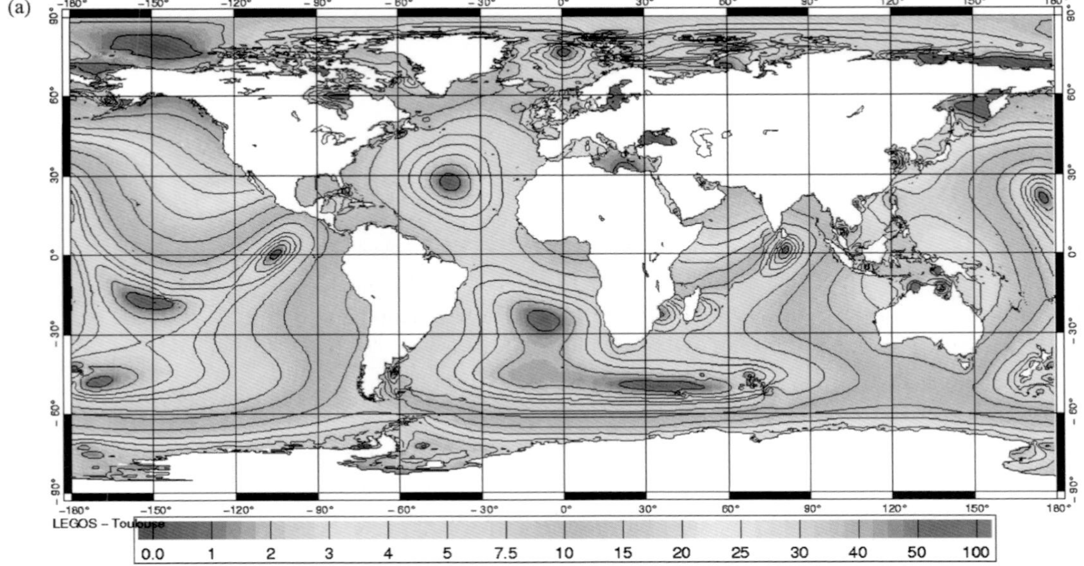

K$_1$ amplitude (cm)

(b)

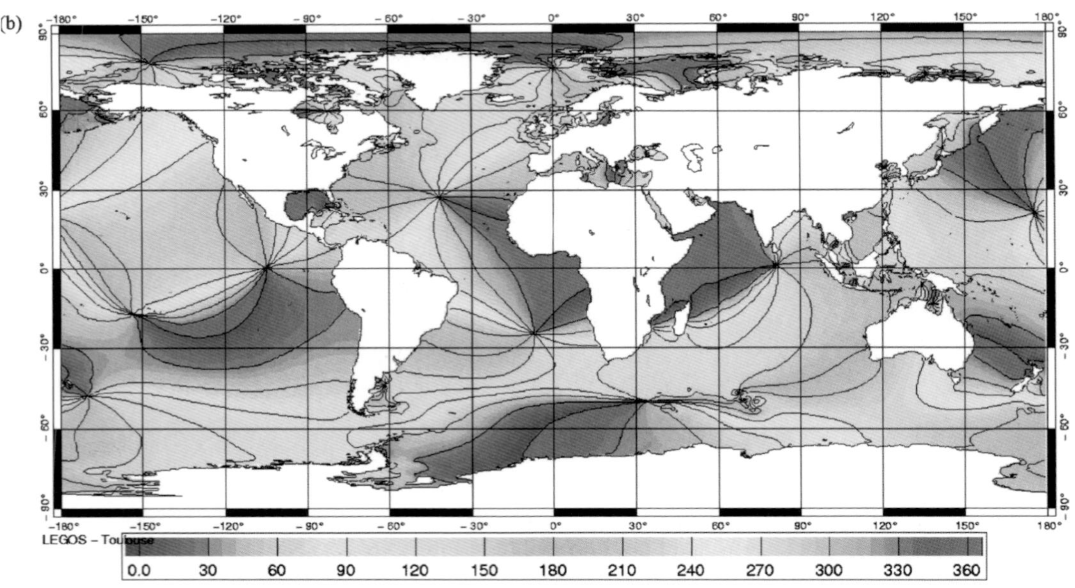

K$_1$ phase, *g* (degrees)

Figure 4.8. A computer-generated map of the **K$_1$** diurnal tidal wave. (a) Amplitudes, which are generally less than those for semidiurnal tides. (b) Phases, with 30° equivalent to two hours. Because the periods and wavelengths are roughly twice those of semidiurnal tides there are fewer amphidromes. Note the Kelvin wave progression from east to west around the coast of Antarctica. (Supplied by Christian Le Provost.) See colour plate section.

Atlantic Ocean rotates anti-clockwise, showing that other types of wave dynamics are also involved.

Similar charts have been prepared for all the other major tidal constituents and each of these contains many amphidromes. The amplitudes of the semidiurnal tides are significantly greater than the amplitudes of the diurnal tides because the oceans have a near-resonant response to forcing at semidiurnal frequencies. The possibility of a near-resonant response for semidiurnal tides is confirmed by rough estimates from Table 4.1. In 4000 m depth, a progressive wave speed of 198 m s^{-1} and a typical ocean length scale of 10 000 km gives a period of 14 hours. For an ocean of 4000 m depth, semidiurnal half-wave resonance needs a width of about 4500 km. The North Atlantic is near to this type of semidiurnal half-wave resonance, and has large semidiurnal tides.

For progressive tidal waves to develop fully as Kelvin waves, the channel in which they propagate must be wide compared with the local Rossby radius (c/f) shown in Table 4.4. At 50° latitude, the Rossby radius in water of 4000 m depth is 1770 km, but this increases as the latitude decreases. The Pacific Ocean at 45° north and south latitude is clearly wide enough for full Kelvin wave development, but the Atlantic Ocean may be too narrow.

In the deep ocean tidal currents are very weak. For a progressive wave of 0.5 m amplitude in 4000 m depth, Equation (4.2) gives a current speed of 0.025 m s^{-1}, which is small compared with currents generated by other forces such as water density differences and wind stresses. Internal tides (see Section 4.5) can have much larger amplitudes and stronger currents, but these currents are mostly irregular and lack coherence.

In the Pacific Ocean, as elsewhere, semidiurnal tides are dominant, but in the North Pacific diurnal tides are also well developed (see Figure 2.1 for San Francisco). The semidiurnal amphidrome at 25°N, 135°W has been firmly established by observations, as has the amphidrome near 25°S, 150°W. This amphidrome gives very small $\mathbf{M_2}$ amplitudes in the vicinity of the Society Islands. Because the $\mathbf{S_2}$ amphidrome is not identically placed, the $\mathbf{S_2}$ tides are dominant in some sites. Where this occurs the semidiurnal tides have high and low water at the same time every day. This is a very local phenomenon, but something similar is also found along the coast of southern Australia from western Tasmania to Cape Leeuwin, where the $\mathbf{M_2}$ and the $\mathbf{S_2}$ constituents are of similar amplitudes. At Thevenard, for example, for part of the spring–neap cycle the times of high waters are the same over several successive days.

Changes in bottom topography have a strong influence on the tidal propagation. As the tide approaches the Hawaiian Islands the wave is diffracted by the sub-marine ridge which extends over 500 km, rather than by the individual islands; this gives a focusing of co-tidal lines from a more extensive region. The semidiurnal tides at Honolulu are unusual because they have a *negative age*, with maximum amplitudes 14 hours ahead of the maximum spring tide gravitational forcing at new moon and full moon, rather than the usual lag of one or two days.

The semidiurnal tides in the northern Indian Ocean have two distinct regimes. The Arabian Sea is broad enough for the standing wave system to develop an amphidrome; however, because this amphidrome is situated close to the equator, it cannot be described simply in terms of Kelvin wave dynamics. To the east, the entrance to the Bay of Bengal is too narrow for an amphidromic system to develop. The tidal wave is propagated to the north along the west coast of Sumatra and Thailand and also along the east coast of Sri Lanka, where the range is low and there is a tendency towards a virtual inland amphidrome. In the south central Indian Ocean there is an extensive region of large semidiurnal tides over which the phases change only slowly. This phenomenon is called an *anti-amphidrome*. A similar semidiurnal system occurs in the south central Pacific Ocean. Large tidal ranges are also observed between the island of Madagascar and the African mainland, in the Mozambique Channel, because a standing wave system develops.

In the Southern Ocean there is an unbroken circumpolar zonal channel for wave propagation around 60°S. Early scientific ideas supposed that this would allow a resonant response to the tidal forces: for a circumference of 20 000 km and a wave travelling in 4000 m water depth, simple theory gives a complete cycle in 28 hours, and allows resonances at harmonics of this period. The observed tides in the Southern Ocean do show the general westward propagation expected for an Equilibrium Tide. This is particularly true for the diurnal tides that have 25-hour periods, but the amplitudes are not sufficiently large to show resonance (Figure 4.8). The narrowness of the Drake Passage between South America and the Grahamland Peninsula of the Antarctic has a strong restricting influence on the circumpolar pattern of wave propagation.

In the central southern part of the Atlantic Ocean there is a complicated amphidromic system that allows a northward progression of the semidiurnal phases along the coasts of Brazil and West Africa. The ranges are relatively large around the equator with phases nearly constant over an extensive area, high water occurring along the whole coast of northern Brazil from 35°W to 60°W within an hour – behaviour consistent with standing wave dynamics.

In the North Atlantic, around 20°N, smaller amplitudes and the rapid northward increase of phase show a tendency for an amphidrome to develop. The most fully developed semidiurnal Atlantic amphidrome is located near 50°N, 39°W. Tidal waves travel around this position in a form which approximates to a Kelvin wave, from Portugal along the edge of the north west European continental shelf towards Iceland and from there west and south past Greenland to Newfoundland. There is considerable leakage of tidal wave energy to the surrounding continental shelves and to the Arctic Ocean, so the wave that is reflected in a southerly direction is much weaker than the wave travelling northwards along the European coast. Subsidiary anti-clockwise amphidromic systems are formed between the Faeroe Islands and Iceland and between Iceland and Greenland. This results in a complete circulation of the phase of the semidiurnal tidal wave around Iceland in a clockwise sense; a similar circulation, also in a clockwise sense, is observed for the semidiurnal tide around New Zealand.

For constituents in the same tidal species, tidal charts in most oceans show broad similarity, but in the Atlantic Ocean there are also significant differences. In the Equilibrium Tide, the ratio between the $\mathbf{M_2}$ and $\mathbf{S_2}$ amplitudes is 0.46; this ratio is much less in the north Atlantic Ocean, and falls to 0.22 at Bermuda. This relative suppression of the principal solar semidiurnal tide $\mathbf{S_2}$ extends over a very large area of the North Atlantic and is observed on both the American and European coasts. It means that the differences between spring and neap tides are much reduced from the (1.46:0.54) ratio of the Equilibrium Tide.

Along the Atlantic coast of North America from Nova Scotia to Florida the ocean tides are nearly in phase, consistent with standing wave dynamics along the northwest to southeast axis; superimposed on this is a slow progression of phase towards the south. The semidiurnal tides of the Gulf of Mexico and the Caribbean have small amplitudes. An anti-clockwise amphidromic system is apparently developed in the Gulf of Mexico for the semidiurnal tides. The diurnal tides in the Gulf of Mexico are larger than the semidiurnal tides because of a local resonant response, but the diurnal tides (Figure 4.8) for the Atlantic Ocean as a whole are generally weaker than in other oceans. Here they can also be described in terms of amphidromic systems, a clockwise system in the south and an anti-clockwise system in the north.

Some marginal basins of oceanic depths, such as the Mediterranean Sea and the Red Sea, which connect to the ocean through narrow entrances, have small tidal ranges. The areas of the basins are too small for direct tidal forces to have much effect, and the areas of the entrance are too restricting for sufficient oceanic tidal energy to enter to build up large tidal amplitudes.

As an example, the small semidiurnal tides of the Red Sea are closely represented by a standing wave having a single central node with a progression in the expected anti-clockwise sense around a central amphidrome between Jedda and Port Sudan (Figure 4.9). In the

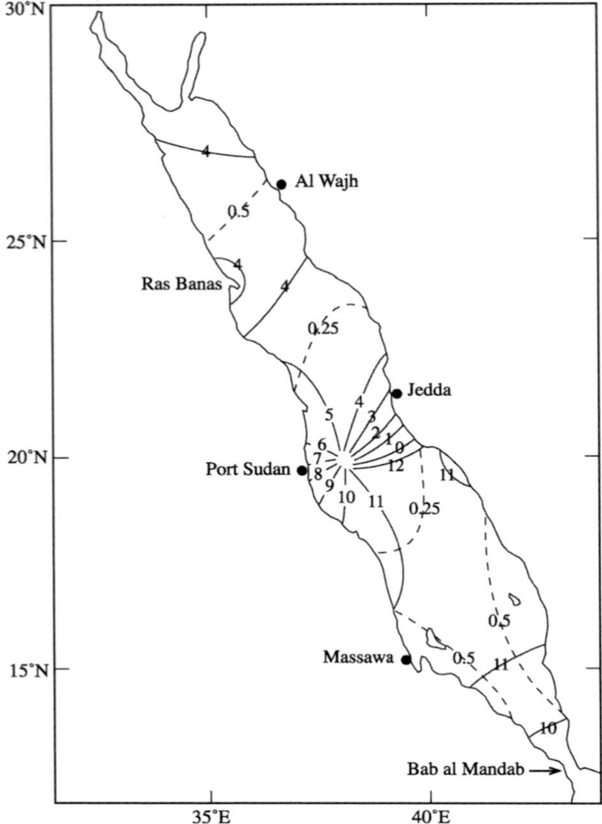

Figure 4.9. A chart of the $\mathbf{M_2}$ tide in the Red Sea. The ranges (dashed curves) are small (shown on the curves, in metres) because tides are driven through the narrow connection to the Indian Ocean at Bab al Mandab. The pattern is very close to that for a half-wave oscillation with an amphidrome between Port Sudan and Jedda.

Mediterranean Sea there are two basins, separated by the Sicilian Channel and the Straits of Messina. Tides of the western basin are strongly influenced by the Atlantic tides which penetrate through the Straits of Gibraltar. Because the connection with the Atlantic Ocean is so restrictive, the influence of direct gravitational forcing within the Mediterranean Sea is probably of comparable importance to the external forcing. In the Adriatic Sea a simple semidiurnal amphidromic system similar to that in the Red Sea is developed. The effect of these standing oscillations is to produce large tides at the northern end, in the vicinity of Venice.

The Arctic Ocean, which has depths in excess of 5000 m, also contains the world's most extensive shallow continental shelf region. Its tides are driven by the Atlantic tides through the connection between Scandinavia and Greenland. As the wave propagates northwards it decreases in amplitude as it circles anti-clockwise around an amphidrome located near 81° 31'N, 130°W, in the deep water of the Canadian Basin.

4.4 Shelf tides

Compared with the deep ocean tides, the patterns of tidal waves on the continental shelf are scaled down and become more variable as the wave speeds are reduced; tidal wave amplitudes are also increased. This wave speed reduction and amplitude increase are shown in Figure 4.2 and Table 4.1 (see also Question 4.5). The Rossby radius, which is also reduced in the same proportion, varies as shown in Table 4.4. For example, the Rossby radius falls from 1770 km in 4000 m depth, to 395 km in 200 m depth, at 50° latitude.

On the continental shelf, tidal waves are strongly influenced by Kelvin wave dynamics and by basin resonances. In very shallow water depths, typically less than 20 m, there will be strong tidal currents and substantial energy losses due to bottom friction. Tidal energy flows onto the shallow regions where it is dissipated. These severe non-linear distortions to the tides are discussed later. In the following discussion it will be possible to describe only a few representative regional cases of shelf tidal behaviour.

The semidiurnal tides of the northwest European continental shelf are shown in Figure 4.10. The Atlantic semidiurnal Kelvin wave travels from south to north. Energy is transmitted across the shelf edge into the Celtic Sea between France and southern Ireland: this wave then propagates into the English Channel where some energy leaks into the southern North Sea. It then spreads into the Irish Sea (see also Section 5.6.2) and the Bristol Channel. The Atlantic wave progresses northwards, taking five hours to travel from the Celtic Sea to the north of Scotland. Here the semidiurnal wave is partly diffracted where it turns to the east and to the south into the North Sea.

The semidiurnal tides of the North Sea, which is broad compared with the local Rossby radius, consists of two complete amphidromic systems and a third, probably degenerate or virtual, system which has its centre in southern Norway. The largest tidal amplitudes occur where the Kelvin south-travelling wave moves along the British coast. Co-range lines are parallel to the coast, whereas co-tidal lines are at right angles, consistent with the Kelvin wave dynamics shown in Figure 4.5. The amphidromic system shown in Figure 4.10 may be compared with the theoretical amphidrome in Figure 4.6 (inverted). Although the southern amphidrome is located near the centre of the sea, progressive weakening of the reflected north-going Kelvin-type wave places the second and third amphidromes further and further to the east of the central axis. Indeed, even the central position of the southern amphidrome is probably partly due to an enhancement of the reflected wave by a north-going wave entering through the Dover Straits.

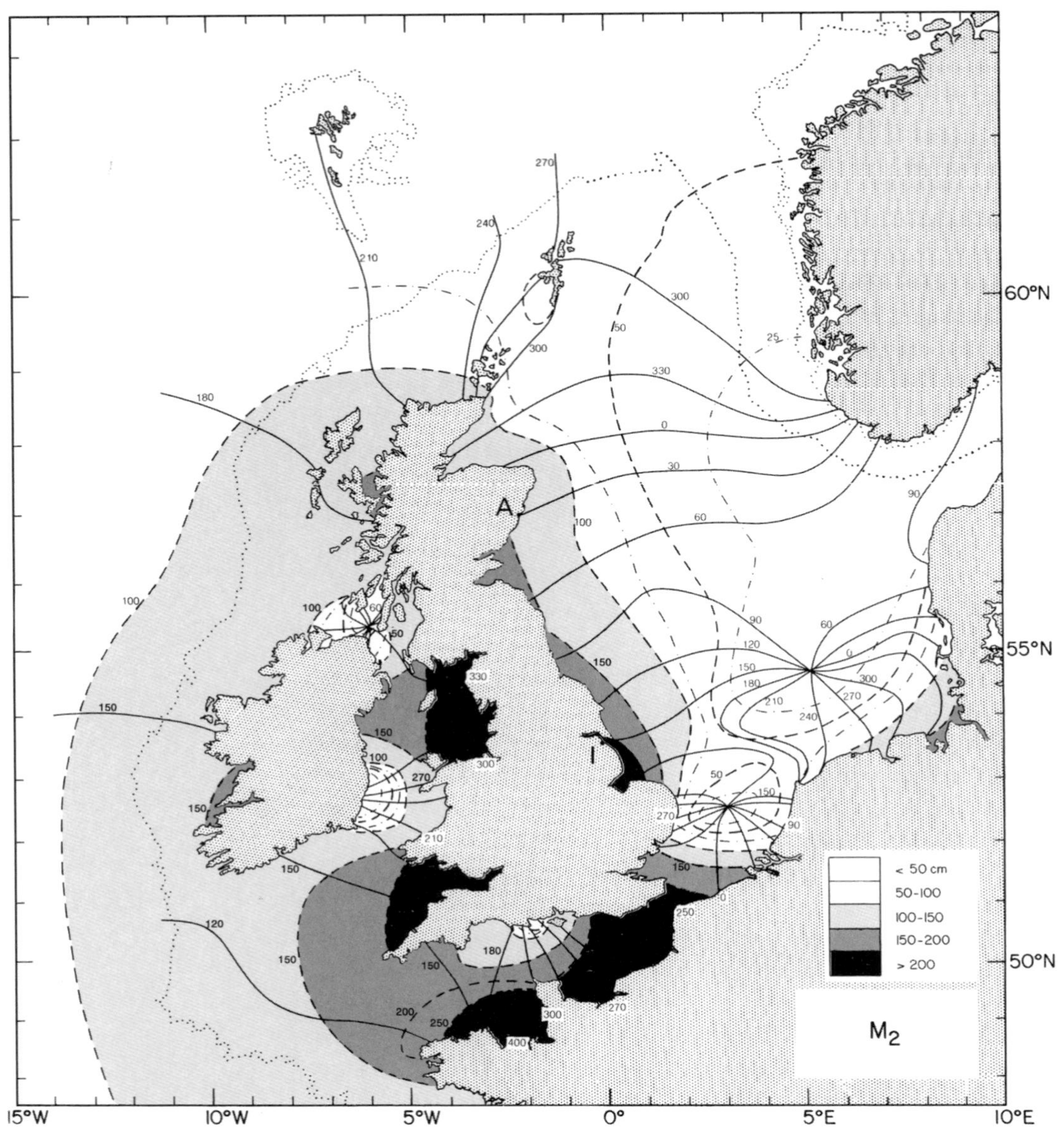

Figure 4.10. The $\mathbf{M_2}$ tides of the northwest European continental shelf plotted from observations of sea level changes and currents. Solid lines are phases in degrees (G); broken lines show amplitudes in centimetres. Note the displacement of the amphidromes in the North Sea away from the centre line because the reflected wave has been weakened by bottom friction. Compare this with Figure 4.6b, inverted. A and I are Aberdeen and Immingham respectively (see Question 6.5). (Crown Copyright.)

The English Channel and the Irish Sea are relatively narrow in terms of the Rossby radius. They respond similarly to an incoming wave from the Celtic Sea. Such a wave takes about seven hours to travel from the shelf edge to the head of the Irish Sea, and a similar time to reach the Dover Straits. The wave which travels along the English Channel reaches the Dover Straits one complete cycle earlier than the wave which has travelled the greater distance around Scotland and through the North Sea. The large tidal amplitudes in the Dover Straits (greater than 2.0 m) are due to these two meeting waves combining to give an anti-amphidrome.

The English Channel tides have a nodal line between the Isle of Wight and Cherbourg. Tidal levels at the Dover Straits have the opposite phase to those at the shelf edge. The large amplitudes on the French coast (3.7 m at St Malo for the $\mathbf{M_2}$ constituent) are due to Kelvin wave dynamics and local standing wave resonance. Because of frictional dissipation and leakage of energy into the southern North Sea, the reflected tidal wave is much weaker than the ingoing wave. A full amphidromic system cannot develop because the Channel is too narrow; instead, there is a clustering of co-tidal lines towards a virtual or degenerate inland amphidrome, located some 25 km inland of the southern English coast.

The semidiurnal tides of the Yellow Sea (Figure 4.11) may be compared with those of the North Sea: the main basin has three amphidromes, progressively displaced from the central axis. The wave which enters from the East China Sea travels in an anti-clockwise sense as a Kelvin wave. The largest amplitudes are found along the coast of Korea. The returning Kelvin wave, which travels south along the coast of China, is much weakened by energy losses to Po Hai Basin and to Korea Bay. As a result, the amphidromes are progressively nearer the Chinese coast and, as in the North Sea, the third amphidrome near 35°N may be virtual.

The nearly simultaneous semidiurnal tides observed over more than 1000 km along the east coast of the United States between Long Island and Florida have been explained in terms of an Atlantic Ocean tide which is nearly simultaneous along the shelf edge. Standing waves develop across the shelf, the coastal amplitude being greatest where the shelf is widest, as shown in Figure 4.12. The smallest range occurs near Cape Hatteras where the shelf is narrowest. Extrapolating the values in Figure 4.12 to a zero shelf width suggests an oceanic $\mathbf{M_2}$ amplitude of about 0.4 m. The shelf is too narrow for any resonant responses to develop.

The semidiurnal tides of Long Island Sound are dominated by a standing wave oscillation with an approach to a quarter-wave resonance which is driven from its eastern entrance. Amplitudes increase from the entrance to the western head of the Sound (Figure 4.13a), and the phases or times are almost simultaneous (Figure 4.13b). Long Island Sound is a body of water of length about 150 km and depth about 20 m. There is also

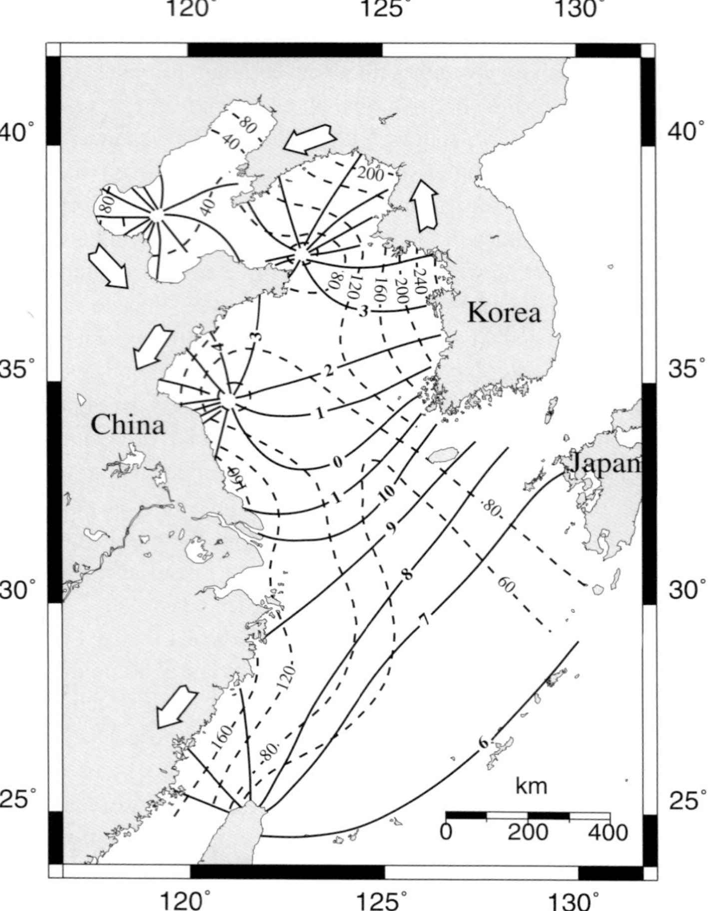

Figure 4.11. The $\mathbf{M_2}$ tides in the Yellow Sea, which show many similarities in amphidrome distribution with the $\mathbf{M_2}$ chart of the North Sea and Figure 4.6b.

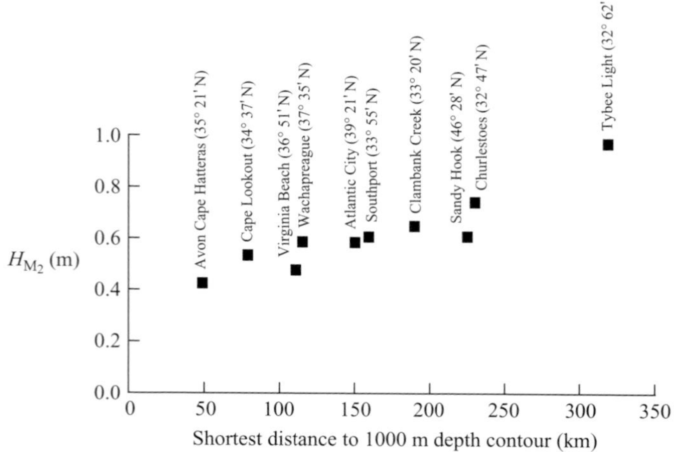

Figure 4.12. The relationship between the width of the continental shelf and the coastal tidal amplitude. Here the $\mathbf{M_2}$ amplitudes for various sites along the coast of the northeast United States are plotted against the shortest distance to the 1000 m depth contour.

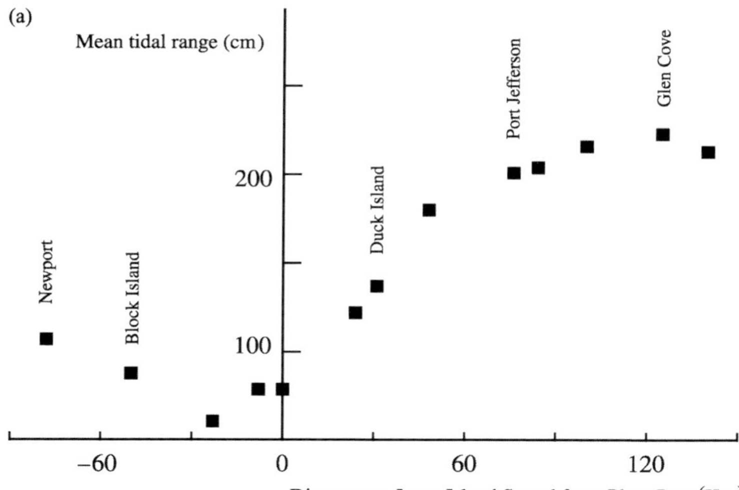

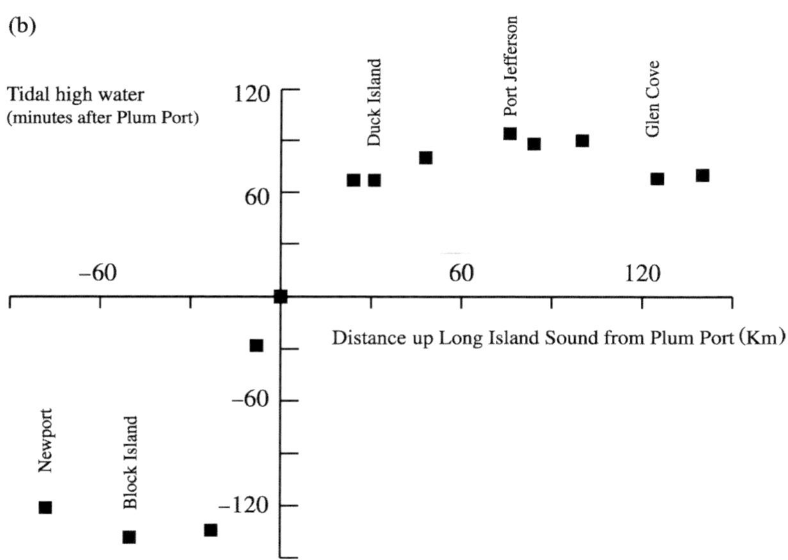

Figure 4.13. The $\mathbf{M_2}$ (a) tidal amplitudes and (b) phases along Long Island Sound, on the east coast of the United States. This can be compared with the standing wave behaviour in Figure 4.4, with maximum amplitudes at the head of the Sound.

a smaller tidal flow which enters through the East River from New York. The effect of this flow from the west into Long Island Sound is to shift the anti-node where maximum tidal amplitudes occur slightly to the east of the head of the Sound. Near to Glencoe the spring tidal range exceeds 2.6 m. The strong tidal currents of New York's East River are driven by the level differences between the Long Island Sound tides at one end and the quite different Atlantic/New York Harbour tides at the other.

Further north, the tidal system that develops in the Gulf of Maine and the Bay of Fundy (see Figure 4.14) consists of near-resonant oscillations

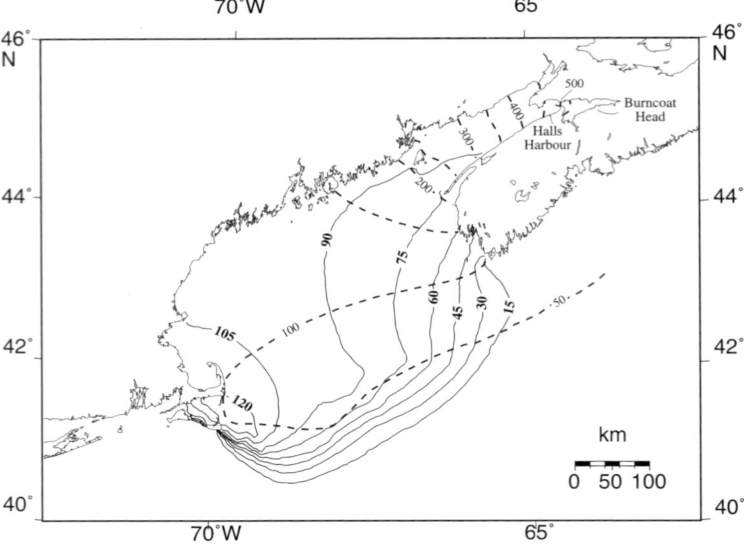

Figure 4.14. A tidal chart of the Gulf of Maine, showing progression to the north and east with maximum ranges in the upper reaches of the Bay of Fundy (based on information supplied by David Greenberg).

which produce one of the world's greatest tidal amplitudes in the Minas Basin (see the box 'The world's biggest tides'). The progressive wave that enters the Gulf from the Atlantic becomes increasingly a standing wave in the upper reaches of the Bay of Fundy. Figure 4.15 shows the difference between high water and low water at Halls Harbour on the Nova Scotia coast. The $\mathbf{M_2}$ amplitudes increase from less than 0.5 m at the shelf edge to more than 5.64 m at Burncoat Head. It has been estimated that the natural quarter-wave period of this system is close to 13.3 hours, which explains the large near-resonant response to semidiurnal tides.

On the east coast of Canada, a full semidiurnal amphidromic system is developed in the Gulf of St Lawrence north of Prince Edward Island, and a further small system appears in Northumberland Strait between Prince Edward Island and New Brunswick. North and west of Prince Edward Island, the tides are mainly diurnal. The tides of the Canadian Arctic are complicated by the many islands and connecting straits. A local resonance produces very large semidiurnal amplitudes in Ungava Bay in northern Quebec (4.36 m for $\mathbf{M_2}$ at Lac Aux Feuilles; again, see the box at the end of this chapter).

The Persian Gulf is a shallow sea with mixed diurnal and semidiurnal tides. It is a largely enclosed basin with only a limited connection to the Indian Ocean through the Strait of Hormuz. Along the major northwest to southeast axis it has a length of about 850 km. The average depth is approximately 50 m, giving it a resonant period near 21 hours. The

Figure 4.15. High and low water on spring tide at Halls Harbour, Nova Scotia, in the Bay of Fundy (supplied by David Greenberg).

Rossby radius at 27°N is 335 km, comparable to the basin width. As a result the response to the diurnal forcing through the Strait of Hormuz is a single half-wave basin oscillation with an anti-clockwise amphidrome. The semidiurnal tides develop two anti-clockwise amphidromic systems, with a node or anti-amphidrome in the middle of the basin. Near the centre of the basin the changes in tidal changes in sea level are predominantly semidiurnal, whereas near the semidiurnal amphidromes they are mainly diurnal. However, near the centre of the basin we have the diurnal amphidrome, and at this node the tidal currents are diurnal.

4.5 Internal tides

This book is mainly about the processes that affect changes in the surface level of the sea, and the most important of these is the astronomical tide. However, there are also tidal movements *within* the interior of the oceans, which have small surface signals and which are of increasing interest to oceanographers. Internal tides (also called baroclinic tides) have been known for a long time, but are difficult to measure. With the advent of long-term current meter and temperature measurements it was observed that internal currents and temperatures often showed tidal variations that in many cases were irregular and not coherent with the movements of the moon and sun, as is the case for surface tides.

From the vertical temperature profiles in the ocean it is possible to convert these temperature variations to equivalent vertical displacements of the water: in many cases these movements are several tens of metres. Ocean currents for the normal tidal waves discussed above can be calculated using Equation (4.2), which shows they are only a few centimetres per second. The observed currents within the ocean at tidal periods are much often larger than this, and so an alternative process must be responsible.

In a simple way we can explain the existence of internal tides by analogy with surface tidal waves. The propagation of waves on the sea surface depends on the density difference between air and water, and on the gravitational acceleration; the air density is small enough to be ignored in the usual theoretical development. Waves may also propagate along density gradients within the ocean, at speeds that depend on the density differences. For internal waves at the interface between two ocean layers we must consider the density of both layers.

For an ocean with two layers of density ρ_1 and ρ_2, and the lower layer (layer 2) much deeper than the thickness of the upper layer D_1, it can be shown that the speed of a long wave over the internal surface is:

$$c = \left[\left(\frac{\rho_2 - \rho_1}{\rho_2}\right) g D_1\right]^{\frac{1}{2}} \tag{4.6}$$

which is the same as for the surface wave, but with a reduced gravity, due to the much reduced density difference required to supply the restoring force. Typical speeds for internal waves are around 1.0 m s^{-1}, much smaller than those of surface waves because of these smaller restoring forces. However, because they can have large amplitudes, the currents associated with the waves can be quite large. The slow speed means that irregular currents and density field variations can change the phase of the tidal currents in unpredictable ways, which explains why internal tides are not usually coherent with astronomical forcing.

An internal tidal wave can change the surface levels very slightly: assuming pressure balance at depth, above the peak of a 10 m internal tidal wave at a density surface where $(\rho_2 - \rho_1)$ is 1 kg m^{-3}, the surface level will be reduced by 0.01 m. These surface changes, though small, can be detected in analyses of long records of altimeter data, but only for internal waves that are regular and astronomically coherent. The method is to analyse repeat tracks over many cycles to remove other effects by averaging, and then to fit amplitudes and phases at successive points along the track for individual constituents. Figure 4.16a shows $\mathbf{M_2}$ amplitude and phase estimates along the TOPEX/Poseidon altimeter tracks across the Hawaiian Ridge, at a 5.75 km spacing, based on three years of data. Internal tidal energy fluxes can be estimated from these amplitudes and phases.

Analyses such as these have shown that energy in the internal tides is radiated from localised areas of marked sudden depth changes and rough topography (e.g. the Hawaiian Ridge in the Pacific Ocean). Other areas include the Mascarene Ridge in the Indian Ocean, Chatham Rise near New Zealand and near seamounts. There is also evidence for strong generation at Straits which separate two areas of deep stratified seas; examples include the Straits of Gibraltar and the Straits of Messina in the Mediterranean Sea. Internal tides are also generated at the shelf edge between the continental shelf and the deep ocean. The internal tides derive their energy from the main tidal currents, where these move density-stratified water over sloping topography. As the tidal currents move up over a rise they move the denser lower water up with them; on the other side the denser water falls back to its normal level, generating internal tidal waves.

At the sea surface the effects of currents can sometimes be detected in satellite images, where they are seen as bands moving away from areas of internal tide generation (Figure 4.16b). The vertical internal movements are important as they bring deep-water nutrient to the surface layers, for example along continental margins and on the coasts of ocean islands, to nourish plankton, coral reefs and seaweeds (see Figure 5.11). The process of internal tidal mixing is enhanced in sub-marine canyons (well-studied examples include the Monterey Bay Canyon off California and the Hudson Canyon off the eastern United States). In these canyons the waves are funnelled and concentrated to give strong flows along the axis of a canyon, converging at the upper end where they meet the continental shelf.

These currents may be responsible for much of the sediment exchange between the shelf and the ocean, but as with many aspects of the dynamics of internal tides, much remains unknown. One sea level variation that can be clearly related to internal tide propagation is the generation of harbour seiches, as we shall see in Chapter 6.

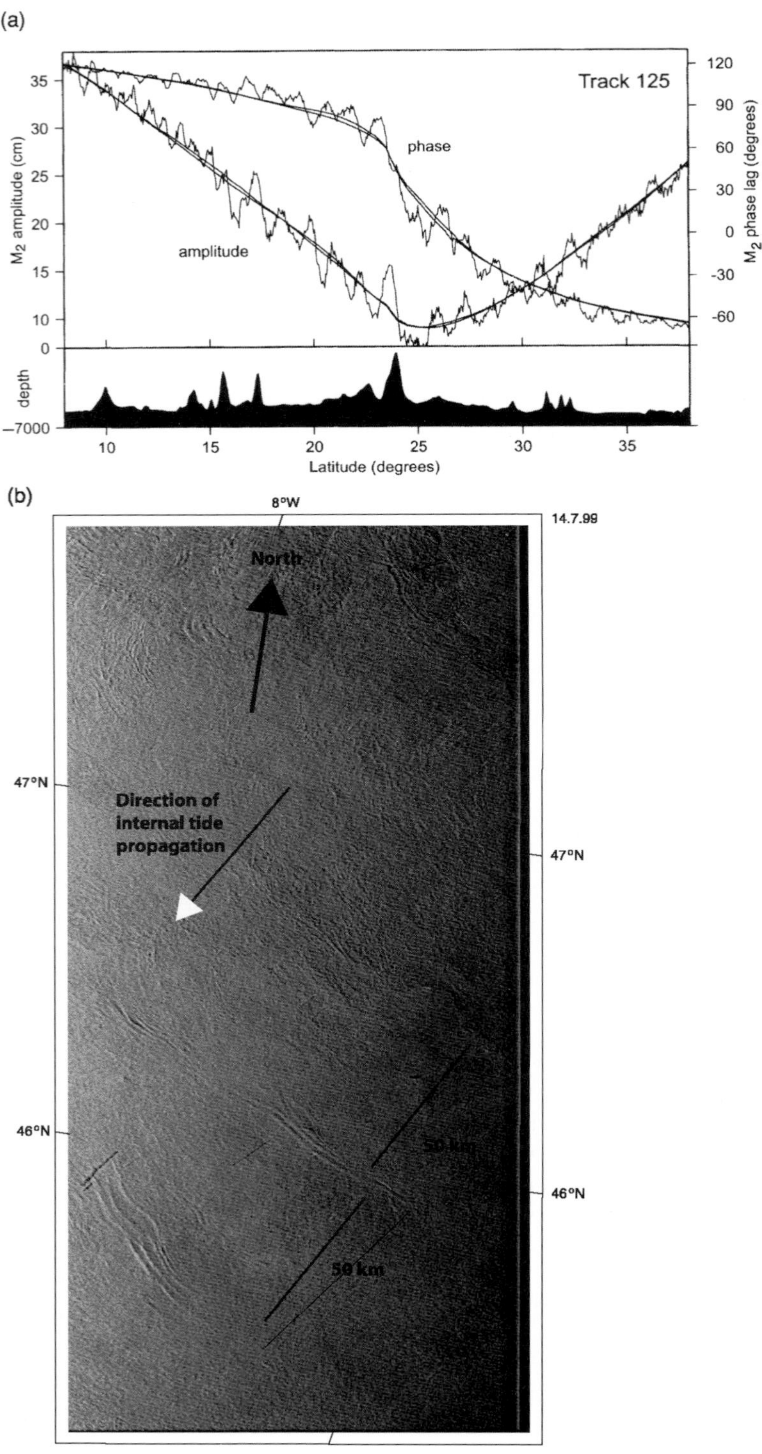

Figure 4.16. (a) Internal tides extracted from TOPEX/Poseidon repeat tracks across the Hawaiian Ridge (from Fu and Cazenave (2001), with permission). (b) Satellite Synthetic Aperture Radar (SAR) image showing internal tides in the Bay of Biscay. The 45 km internal tide wavelength spacing between tidal maximum internal wave activity is as expected for propagation along the thermocline density interface (supplied by Adrian New).

4.6 Tidal energy

The energy lost by the tides through bottom friction in shallow water is converted to turbulence of decreasing scale, which eventually generates a small amount of heat. The original source of this energy is the dynamics of the earth–moon system and, over geological time, gradual but fundamental changes have occurred because of these steady losses of energy. The slowing down in the rate of rotation of the earth is increasing the length of the day by 1 s in 41 000 years. The angular momentum of the earth–moon system is conserved by the moon moving away from the earth at a rate of 37 mm yr^{-1}. The total rate of tidal dissipation due to the $\mathbf{M_2}$ tide can be calculated rather exactly from the astronomical observations at 2.50 ± 0.05 TW (1 TW $= 10^{12}$ W), of which 0.1 TW is dissipated in the solid earth. The total lunar dissipation is 3.0 TW, and the total due to both sun and moon is 4.0 TW. For comparison, geothermal cooling causes a heat loss of 30 TW. Solar radiation energy input to the earth is five orders of magnitude greater than the tidal energy. The 1999 total installed electrical capacity for the United Kingdom was 70 GW (70×10^9 W) and the average electrical energy consumption for New York State was 18 GW.

To gain a better understanding of the processes of energy loss we must first look at how the drag of the sea bed affects tidal currents. Bottom friction is approximately proportional to the square of the speed of the current which it opposes. From this it can be shown that the average energy dissipation over a tidal cycle is proportional to the cube of the current speed (this discussion links very closely with Section 5.6.1). This cubic law of energy dissipation means that tidal energy lost to bottom friction is a phenomenon strictly localised to a few parts of the world where there are strong currents. It also means that much more energy is dissipated at spring tides than at neaps. In the Equilibrium case the rate of tidal energy dissipation at springs $(\mathbf{M_2} + \mathbf{S_2})^3$ is twenty times the rate of dissipation at neaps $(\mathbf{M_2} - \mathbf{S_2})^3$. Notable among these local areas of intense energy dissipation are the northwest European Shelf, the Patagonian Shelf, the Yellow Sea, the Timor and Arafura Seas, Hudson Bay, Baffin Bay and the Amazon Shelf.

It now appears that around 25 per cent (1 TW) of the tidal energy may be dissipated by internal tidal waves in the deep ocean, where the dissipation processes contribute to vertical mixing and the breakdown of stratification. Energy losses may be concentrated in a few areas, for example where the topography of mid-ocean ridges and island arcs create favourable conditions, as we discussed in Section 4.5.

Are tides changing?

The tides are driven by the gravitational forces of the moon and sun, whose movements are extremely regular and stable. Ocean tidal amplitudes depend on the response of the oceans to the tidal forces which in turn change as the depth and shape of the ocean basins change. These will change only slowly over geological time. Direct comparisons of old and recent tidal observations are rare, but Cartwright (1971) found that between 1761 and 1961 the oceanic semidiurnal tides at St Helena in the South Atlantic were constant in amplitude to within 2 per cent. For the French port of Brest, which is well connected to the Atlantic Ocean, Cartwright (1972) found similar stability in the tidal amplitude between 1711 and 1936. Between 1842 and 1979 the amplitudes of the tides at Castletown and Courtown (Figure 2.1) on thesouth west and east coasts of Ireland had increased by less than 2 per cent and the phases were constant to within one degree, the equivalent of two minutes.

Nevertheless, locally there have been significant changes in tidal amplitudes. In London, more than 80 km up the River Thames from the North Sea, high water levels have increased by around 0.8 m per century whereas low water increases were only around 0.1 m per century; similar increases have been found in northern Germany. The increase in range is found to be due to an increase in high water levels while the low water levels have remained roughly the same. The standard deviation of sea levels about the annual MSL, shown in Figure 3.2, includes this small trend against the normal 18.6-year nodal variations. Along the Netherlands and Belgian coasts the tidal range has also increased locally: at Flushing the increase in mean tidal range over the period 1900–80 was 0.14 m per century, or approximately 4 per cent.

For St John Harbor in the Bay of Fundy, Godin (1992) found an amplification of $\mathbf{M_2}$ of around 4 per cent per century, but it appeared that $\mathbf{S_2}$ was decreasing. For tidal systems near resonance, as in the Bay of Fundy, there will be greater sensitivity to changes in depth and local frictional losses of energy through bottom friction. The evidence that tidal ranges in the Bay of Fundy were as small as 2 m only 10 000 years ago will be discussed in Chapter 9. Natural processes and engineering works can both affect local tides. Local siltation, changes in dredging practices for navigation, and canalisation of rivers are all relevant factors.

The world's biggest tides

Where are the world's biggest tides? This is a simple, often asked question, but there is no simple answer. It depends on what is meant by 'biggest tide'. If we want the largest $\mathbf{M_2}$ tides, these are found in the Minas Basin at the upper end of the Bay of Fundy. Near Burncoat Head (Figure 4.17a) the $\mathbf{M_2}$ amplitude is 5.64 m. Large values of $\mathbf{M_2}$

Figure 4.17. (a) A view of Burncoat Head at low tide. The world's largest $\mathbf{M_2}$ tides occur here. (b) Clevedon, on the coast of the Bristol Channel (see Figure 5.6), United Kingdom, makes a more modest claim.

are also found at Hansport and other places in the Minas Basin. This means that averaged over many tides the tidal range is greatest in the Minas Basin. A different measure of 'biggest tide' is the highest high water sea level above the mean sea level generated by the tides; these are probably found at the head of the Minas Basin, around Truro, but here the total *range* is smaller because the shore dries out as the tides recede and extreme low waters are not observed.

Interestingly, in the Minas Basin the semidiurnal solar tide $\mathbf{S_2}$ is only about 15 per cent of $\mathbf{M_2}$, and so the spring–neap modulations are not as large as in the Equilibrium Tide (see Section 4.4). This means that there are sometimes larger tidal ranges at other places where $\mathbf{S_2}$ is larger. The tides in Lac Aux Feuilles in Ungava Bay, in the Hudson Strait of northeast Canada also has exceptionally high tides ($\mathbf{M_2}$ = 4.36 m); but here $\mathbf{S_2}$ is bigger than in the Bay of Fundy, so there are larger spring–neap modulations. For 30 March 2002, for example, the tidal range in Lac des Feuilles, calculated from the Canadian tide tables (see Table 3.4) was 16.0 m – slightly larger than the range for Burncoat Head.

Further reading

For a very readable account at a general level, I recommend Garrett and Maas (1993); if you can find a copy, Defant (1958) is excellent. The theory of wave dynamics and of long waves on a rotating earth is covered in many oceanographic textbooks. The more accessible of these include Von Arx (1962), Neumann and Pierson (1966), Pond and Pickard (1995), Knauss (1997), Mooers (1999) and the Open University (2000). More rigorous and comprehensive accounts are given in Lamb (1932), Defant (1961) and Gill (1982). Parker (1991) includes several valuable review chapters.

Descriptions of ocean tides are found in many hydrographic and scientific publications. A good start is the tidal chapter by Christian Le Provost in Fu and Cazenave (2001); see also Le Provost *et al.* (1998). Defant (1961) is still probably the most comprehensive description of the tidal dynamics of individual seas around the world. Accounts of local tides abound often in older, sometimes obscure publications. These include Dawson (1920), Marmer (1926), Redfield (1980) and Dohler (1986). For UK waters, Admiralty Chart 5058 has recently been revised; it shows co-tidal and co-range lines for spring tides ($\mathbf{M_2} + \mathbf{S_2}$). Shelf tides are also reviewed in Simpson (1998) which includes aspects of tidal distortion that we will consider in the next chapter.

Questions

4.1 Calculate the natural period for a forced semidiurnal quarter-wave oscillation in Long Island Sound, assuming a depth of 20 m and a length of 150 km. Comment on the value you get.

4.2 In the Equilibrium Tide $\mathbf{S_2} = 0.46\mathbf{M_2}$. In the North Atlantic Ocean $\mathbf{S_2}$ is suppressed so that, typically, $\mathbf{S_2} = 0.33\mathbf{M_2}$. Assuming that the energy dissipated by tidal currents is proportional to the cube of the current speed, what is the ratio of spring tide to neap tide rates of energy dissipation for (a) the Equilibrium Tide and (b) the North Atlantic tides? Assume that tidal currents are proportional to tidal amplitudes.

4.3 At Newlyn and at some ports along the north German coast, low water levels have remained the same but high water levels have increased slightly over the past 100 years. Can you explain this?

4.4 By comparing Figure 4.4 with Figure 4.13, discuss how the tides of Long Island Sound are different from perfect quarter-wave tidal resonance.

4.5 In Figure 4.2, if a semidiurnal wave travels from the ocean ($D =$ 4000 m) onto a shallow continental shelf ($D =$ 100 m), what is the difference of wave speed? Would the result be different for a diurnal tidal wave? What are the tidal wavelengths on the shelf for semidiurnal and diurnal cases?

4.6 Can you find a real semidiurnal amphidrome in Figure 4.10, not in the North Sea?

4.7 Use Equation (4.6) to calculate the speed and wavelength of a semidiurnal internal tidal wave propagating on the main thermocline at a depth of 5400 m; the density above the thermocline is 1026 kg m^{-3} and that below is 1027 kg m^{-3}. Locate these length and time scales on Figure 1.3.

Chapter 5
Tides near the coast

In Chapter 4 we showed how the amplitudes of the tidal waves generated in the deep oceans increase when they spread onto the shallow surrounding continental shelves. In these shallower waters other processes, including standing wave generation and local resonances, alter the characteristics of the tidal waves. In this chapter we will consider the further and more extreme distortions that occur as the tidal waves propagate into the even shallower coastal waters and the rivers. The behaviour of these distorted tides is very important for near-shore human activities such as recreational pursuits and coastal navigation. The distortions are also important for the geological and biological processes in the coastal zone, which we will explore further in Chapter 9.

Firstly, we consider the physical factors that cause distortion and then we show how extra tidal constituents at higher frequencies can fit these distorted tides. Although this book is not looking at currents in detail, it is appropriate to discuss in a general way how tidal sea level oscillations can control flow through narrow channels. We then look at how tidal distortions become more extreme as the wave travels into estuaries and up rivers, leading in some cases to the ultimate distortion, the tidal bore. Finally we will look again at how tidal energy is lost and how this affects the mixing and stratification of shallow waters.

Before looking at these distortions it may be helpful to remind ourselves about the nature of non-linear physical processes. Non-linear is a technical term, but it has a clear physical meaning. When we considered response analyses of tides in Chapter 3, there was an input to a 'black box' and an output from the box. In a linear system the output is proportional to the input: if the input is doubled then so is the output. Ocean tides largely respond to astronomical forcing in this way. If a system is non-linear, then the output may change in proportion to the square or other powers of the input. Many physical processes that affect tides in shallow water are non-linear. For example, the frictional drag on a tidal current increases as the square of the current speed, so it increases fourfold if the current speed doubles. In the following discussion we will look at the physics of some of these processes in a quantitative way, but we will not go into the theoretical developments in detail.

5.1 Hydrodynamic distortions

In this section we consider three separate physical factors, each of which may contribute to tidal distortions. Firstly, the stronger currents that develop in the shallow waters are resisted by the drag due to bottom friction, a process that eventually removes much of the propagating tidal energy and reduces the wave amplitudes. Secondly, the amplitude of the tidal waves becomes a significant fraction of the total water depth and this restricts the wave's travel. A third distorting factor is the influence of topography: irregular coastlines and varying depths impose complicated tidal current patterns; where a current has a curved path, there must be a surface gradient at right angles to the flow to provide the force needed to change the direction.

In practice the relationships are more complicated because there are several types of non-linear activity that may interact and influence the dynamics. It is not always possible to be sure which effects are more important in particular cases. However, it is useful to start with a description of these three effects individually.

5.1.1 Bottom friction

The effects of bottom friction are to oppose the tidal current flow and to remove energy from the tidal motion. Experiments and theoretical arguments show a relationship between the current speed and the drag:

$$\text{Drag} = C_{\text{D}}\rho u^2 \quad \text{per unit area} \tag{5.1}$$

where u is the current speed, C_{D} is a dimensionless drag coefficient and ρ is the water density. The drag is always opposing the current. This is a non-linear process and the relationship involving u^2 is sometimes called the squared law of bottom friction. The value of C_{D} depends on the level above the sea bed at which the current is measured; for calculating forces on the sea bed, this level is usually taken as 1 m, and in this case the value of C_{D} may lie between 0.0015 and 0.0025.

Figure 5.1. Distortion of a progressive wave travelling in shallow water up a channel in the positive *x*-direction. Note the shorter rise time as the wave passes a fixed observer, as the distortion increases.

Suppose that the tidal currents vary harmonically, and the flow up and down a narrow channel is given by $u = U_{M_2} \cos 2\omega_1 t$. The u^2 term in Equation (5.1) may then be expanded mathematically (allowing for the fact that the frictional drag reverses direction when the current reverses) as a series that contains only odd harmonics of $\cos 2\omega_1 t$:

$$u|u| = U_{M_2}^2 (a_1 \cos 2\omega_1 t + a_3 \cos 6\omega_1 t + a_5 \cos 10\omega_1 t + \cdots) \quad (5.2)$$

where $a_1 = 8/3\pi$, $a_3 = 8/15\pi$ etc. This shows that $\mathbf{M_2}$ tidal currents in a channel will result in frictional forces that include the term in $\cos 6\omega_1 t$, which is represented in harmonic constituent terms by $\mathbf{M_6}$. In the more general case of two-dimensional flow, both even and odd harmonics of the basic tidal frequencies are present in the frictional resistance terms.

5.1.2 Finite water depth

Most people are familiar with the behaviour of wind waves as they approach the shore. Gradually the wave front steepens until eventually the wave breaks. Similar steepening occurs for tidal waves, but the initial distortion is not obvious to the casual observer because the wavelengths and periods are so much greater. The shallowing is no longer essential for the wave to steepen; the essential requirement is that the wave amplitude is not small compared with the total depth. Figure 5.1 shows an exaggerated profile of a wave moving in the direction of increasing x. At any fixed location the rise and fall of the water as the wave passes will not take equal times. The rate of rise of the water level is more rapid than the rate of fall. This difference becomes greater as the wave progresses (x increases).

It is easy to see in a general way why the crests of the waves tend to catch up with the troughs. Previously (Section 4.2.1) we have shown that the wave speed (c) is given by $c = (gD)^{\frac{1}{2}}$ where the depth is large compared with the amplitude and the wavelength is long compared with the depth. Since the wave speed decreases as the water depth decreases, the troughs of the waves will tend to be overtaken by the crests, which are travelling in deeper water. This gives rise to the asymmetry. In fact, more detailed theory shows that where the wave amplitude ζ is comparable to the total depth D a more exact relationship for the wave speed is

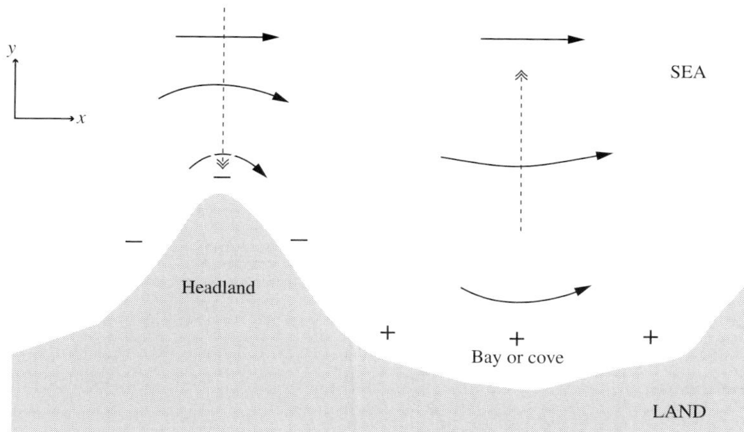

Figure 5.2. Curvature of streamlines for tidal flows near a coast (solid arrows). The surface gradients, which produce the curvature, are represented by the broken arrows, which point down the slope. The currents are reversed half a tidal cycle later, but the gradients are the same.

given by $c = [g(D + \frac{3}{2}\zeta)]^{\frac{1}{2}}$. The differences between the wave crest and wave trough speeds given by these more elaborate formulae increase the steepening of the wave even more.

In the simple case of an $\mathbf{M_2}$ tidal wave, we can show that an $\mathbf{M_4}$ component is generated as the wave progresses into shallow water. If the $\mathbf{M_2}$ tidal wave $\zeta = H_{\mathrm{M_2}} \cos 2\omega_1 t$ travels in a channel of depth D, mathematical analysis for the form of the wave in the channel has as the first two terms:

$$\zeta = H_{\mathrm{M_2}} \cos(kx - 2\omega_1 t) - \frac{3}{4}\frac{kxH_{\mathrm{M_2}}^2}{D} \sin 2(kx - 2\omega_1 t) + \cdots \quad (5.3)$$

Here, the distance along the channel is x and k is the inverse of the wavelength (or the fraction of a wavelength per metre). At the start of the channel where $x = 0$, the second term is zero, and we have the initial $\mathbf{M_2}$ tidal wave we defined above.

The second term is a harmonic $\mathbf{M_4}$ of the original wave. Its amplitude increases linearly as the distance increases along the channel. The amplitude is also bigger if the channel depth is small, and it increases in proportion to the square of the amplitude of the original wave. This is another example of non-linear interaction. Further refinement leads to solutions containing other higher harmonics of $\mathbf{M_2}$.

5.1.3 Flow curvature

Consider a tidal current flowing along an irregular coast of headlands and indentations (bays and coves). The tidal currents tend to follow the contours of the coast, deflecting into the bays and out again around the headlands. The forces that change the directions of current flow along the coast are given by the pressure gradients across the directions of the streamlines. In Figure 5.2 the arrows on the broken lines show the downward slope of the sea surface to give this gradient. Assuming that the sea surface well away from the coastal influences is a horizontal level surface, then the sea levels at the headland will be slightly lower, and the levels in the bays will be slightly higher than those offshore.

These gradients are in the same sense, upwards or downwards, when the direction of flow is reversed half a tidal cycle later. Of course, because this process is not related to the earth's rotation it is the same in both hemispheres. Therefore, there is a relative depression of sea level at the headlands for both the flood and the ebb flows, that is, twice in a tidal cycle. At the coast, in a harmonic analysis of the headland sea levels, this appears as an enhanced $\mathbf{M_4}$ constituent. There is also a small long-term depression of the mean sea level at the headland. Because headlands are sharper coastal features than bays, the flow curvature and associated surface gradients there are more severe. As a result, gauges located in broad bays will measure levels that are more representative of open-sea conditions than those located near headlands.

The gradient that is necessary to produce the flow curvature can be calculated in terms of the current speed u and the radius of curvature of the flow (the same physics applies for both the forces which give this cross-track acceleration of the current and those that keep the moon or sun (or a satellite) in orbit, as discussed in Section 2.2):

$$g \times \text{surface slope} = -\frac{u^2}{r} \tag{5.4}$$

where r is the radius of the curvature of the flow. For a radius of curvature of 1 km and a current speed of 1 m s^{-1}, the gradient is 0.1 m in 1 km. The effects of this on sea level at the coast will depend on the distance over which this curvature of flow and the corresponding surface slope gradient are maintained. Around Portland Bill, which protrudes into the English Channel, this curvature effect is responsible for an $\mathbf{M_4}$ amplitude of 0.10 m.

5.2 Representation by higher harmonics

We have shown in a general way how the progression of a tidal wave in shallow water is modified by bottom friction and other physical processes that depend on the square or higher powers of the tidal amplitude itself. Figure 5.3 shows the presence of extremely strong distortions at Southampton on the south coast of England; the double high and double low waters are due to non-linear shallow-water effects. As for the main astronomical tides, these distortions can also be expressed as simple harmonic constituents. These new constituents have angular speeds that are multiples, sums or differences of the speeds of the

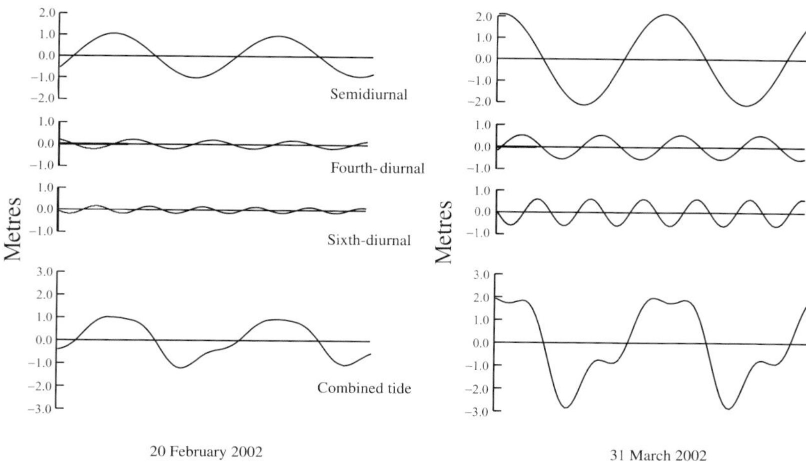

Figure 5.3. Showing the effects of shallow-water distortion on the tides at Southampton, for neap tides (20 February) and spring tides (31 March). Double high waters occur only on larger tides.

astronomical constituents listed in Table 3.2. Figure 5.3 shows how harmonic terms in the fourth-diurnal and sixth-diurnal species can fit and reproduce the effects, for both neap tides (20 February) and spring tides (31 March).

To illustrate this generation of higher harmonics in a different way, consider that the non-linear shallow-water effects are proportional to the square of the tidal water level through a spring–neap cycle. This means that they are proportional to:

$$\{H_{\mathrm{M}_2} \cos 2\omega_1 t + H_{\mathrm{S}_2} \cos 2\omega_0 t\}^2$$

which expands to:

$$\begin{aligned}\Big\{&\tfrac{1}{2}\left(H_{\mathrm{M}_2}^2 + H_{\mathrm{S}_2}^2\right) + \tfrac{1}{2}\left(H_{\mathrm{M}_2}\right)^2 \cos 4\omega_1 t + \tfrac{1}{2}\left(H_{\mathrm{S}_2}\right)^2 \cos 4\omega_0 t \\ &+ H_{\mathrm{M}_2} H_{\mathrm{S}_2} \cos 2\left(\omega_1 + \omega_0\right) t + H_{\mathrm{M}_2} H_{\mathrm{S}_2} \cos 2\left(\omega_0 - \omega_1\right) t\Big\}\end{aligned} \tag{5.5}$$

This new relationship has produced additional harmonics at $4\omega_1$ and $4\omega_0$ that are called $\mathbf{M_4}$ and $\mathbf{S_4}$ because their frequencies appear in the fourth-diurnal species. The term at speed $2(\omega_1 + \omega_0)$, also in the fourth-diurnal species, is called $\mathbf{MS_4}$ to show that it originates from a combination of $\mathbf{M_2}$ and $\mathbf{S_2}$. A long-period harmonic constituent is also produced. This has a speed equal to the difference in the speeds of the interacting constituents, $2(\omega_0 - \omega_1)$, called $\mathbf{MS_f}$, which has a period of 14.78 days, exactly the same as the period of the spring–neap modulations of the tidal amplitudes. The interaction has also produced a change in the mean sea level, represented by the terms in $H_{\mathrm{M}_2}^2$ and $H_{\mathrm{S}_2}^2$.

Table 5.1. *Some of the more important higher tidal harmonics generated in shallow water.*

	Generated by	Angular speed	(degrees per hour)
Long-period			
MS_f	M_2, S_2	$S_2 - M_2$	1.0 159
Semidiurnal			
$2MS_2^*$	M_2, S_2	$2M_2 - S_2$	27.9 682
Fourth-diurnal			
MN_4	M_2, N_2	$M_2 + N_2$	57.4 238
M_4	M_2	$M_2 + M_2$	57.9 682
MS_4	M_2, S_2	$M_2 + S_2$	58.9 841
Sixth-diurnal			
M_6	M_2	$M_2 + M_2 + M_2$	86.9 523
$2MS_6$	M_2, S_2	$2M_2 + S_2$	87.9 682
Eighth-diurnal			
M_8	M_2	$M_2 + M_2 + M_2 + M_2$	115.9 364

*Also contains a significant gravitational component (**$2MS_2$** has the same period/speed as μ_2).

In practice a whole range of extra constituents are necessary to represent distortions in shallow water. Some of the more important of these are given in Table 5.1. For a full analysis and prediction of tides in shallow-water regions some authorities use more than 100 harmonic constituents.

5.3 Southampton tides

The non-linear nature of the distortions means that their effects are much stronger at spring tides than at neaps. In this section we look at the way in which the higher harmonics grow as the tidal range increases; we illustrate this with an extreme example, the Southampton tides, which we looked at briefly in the previous section.

In a few shallow-water areas, the distortion takes the form of a double high water or a double low water. Double high waters are especially notable at Southampton. Elsewhere, double high waters are also observed at Falmouth in Buzzards Bay, Massachusetts, USA and at the Hook of Holland and Den Helder on the Dutch coast. There are also double low waters at Marion in Buzzards Bay.

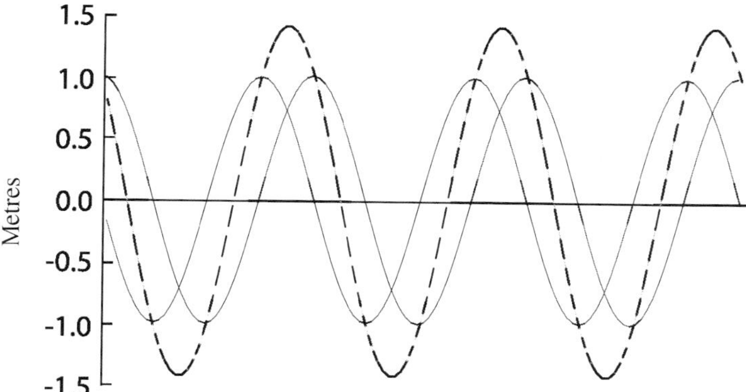

Figure 5.4. Showing how adding two $\mathbf{M_2}$ tides of different phases, equivalent to a three-hour delay, produces a new $\mathbf{M_2}$ tide (broken curve) of intermediate phase, but no harmonic distortions.

Look again at Figure 5.3. It shows the tidal variations at Southampton for both spring and neap tides. During spring tides, following low water the water level rises, but there is then a slackening of the tidal stream and a water level stand for a further two hours before the final rapid rise to high water, over the next three hours. The slackening effect is known locally as the 'young flood stand'. The flood and the double high water last approximately nine hours, leaving only three hours for the tidal ebb, which therefore has very strong ebb currents. This is quite different from the usual pattern of tidal behaviour in estuaries, as we shall see in Section 5.5.

Popular explanations of the double high water in terms of the travel of two separate $\mathbf{M_2}$ progressive waves around the Isle of Wight (to the south of the entrance to Southampton Water) are wrong: Newton was among the earliest to point out that, from basic trigonometry, adding two harmonic waves that have the same period but different phases produces a single bigger combined wave of the same period and of intermediate phase. Figure 5.4 confirms that no extra maxima or minima are created by this superposition.

We can investigate the Southampton double high waters in more detail by examining how the shallow-water higher harmonic tides vary from day to day as the semidiurnal tides change. In this region the diurnal tides are very small and can be ignored for our purpose.

An alternative to the analysis of long periods of data for several harmonics, which are separated from each other by frequency increments equivalent to a month (tidal *groups*) or a year (tidal *constituents*), is to analyse day by day for the amplitude and phase of the harmonic present on that day, in each *species*. These daily harmonics may be called $\mathbf{D_1}$, $\mathbf{D_2}$, $\mathbf{D_4}$, etc. by analogy with the usual notation for the naming of harmonic constituents. Modulations of these $\mathbf{D}_n$ terms will take place over the spring–neap cycle, over a month and over a year. This is the same as the

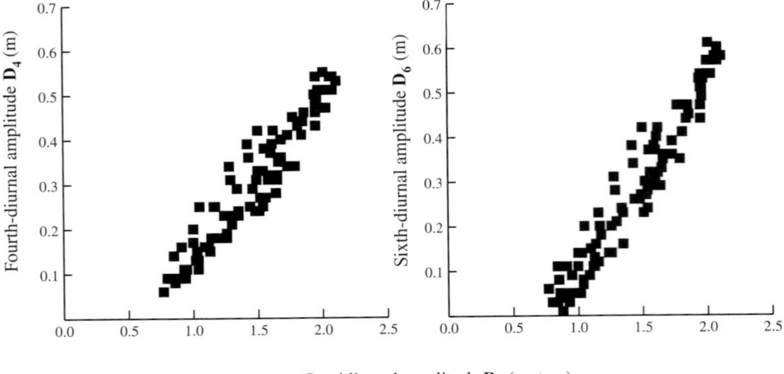

Figure 5.5. The amplitude of the *daily* non-linear tides, $\mathbf{D_4}$ and $\mathbf{D_6}$, as a function of the amplitude of the composite semidiurnal tide $\mathbf{D_2}$, also determined each day, at Southampton, February–April 2002.

way the many lunar constituents in the more detailed one-year analyses vary over 18.6 years (see Chapter 3). For example $\mathbf{D_4}$ is a composite of all the fourth-diurnal harmonic tidal constituents.

Although not suitable for prediction purposes, daily analysis can give an insight into the physics of non-linear dynamics that is not possible by other means. Figure 5.5 shows the variations of the $\mathbf{D_4}$ and $\mathbf{D_6}$ amplitudes at Southampton, as a function of the $\mathbf{D_2}$ amplitude. Both are very small indeed when the tidal amplitudes are less than 1.0 m. The $\mathbf{D_6}$ amplitudes increase from near zero on neap tides to more than 0.6 m on spring tides. The $\mathbf{D_4}$ amplitude increases from about 0.05 m on neap tides to 0.55 m on spring tides. As the tidal amplitude $\mathbf{D_2}$ increases, $\mathbf{D_4}$ increases rapidly; $\mathbf{D_6}$ increases even more rapidly as would be expected for a higher power law of interaction. Section 5.2 has shown that a combination of $\mathbf{M_2}$ and $\mathbf{S_2}$ leads to $\mathbf{D_4}$ enhancement through terms in $\mathbf{M_4}$, $\mathbf{MS_4}$, etc., if a squared law of interaction applies. Further higher harmonics such as $\mathbf{D_6}$ are generated by higher powers of interaction. The very rapid increase in $\mathbf{D_6}$ may imply some kind of local resonance, perhaps in the form of cross-channel oscillations.

It can be shown that a double high water can only be generated if H_{D_4} is at least a quarter of H_{D_2}. Similarly, for a double high water generated by sixth-diurnal harmonics, H_{D_6} must be more than one-ninth of H_{D_2}. At Southampton neither $\mathbf{M_4}$ nor $\mathbf{M_6}$ is large enough on its own to produce double high water in $\mathbf{M_2}$; rather, it is a combination of the several shallow-water harmonic terms that represent the observed distortions at spring tides. These shallow-water distortions are not big enough to give double high waters on neap tides (see Question 5.3).

The growth of the higher harmonics as $\mathbf{D_2}$ increases is generally less spectacular at other ports, but will follow the same pattern. Observations at other places along the coast of the English Channel show that the extreme shallow-water distortions are more extensive and that the effects

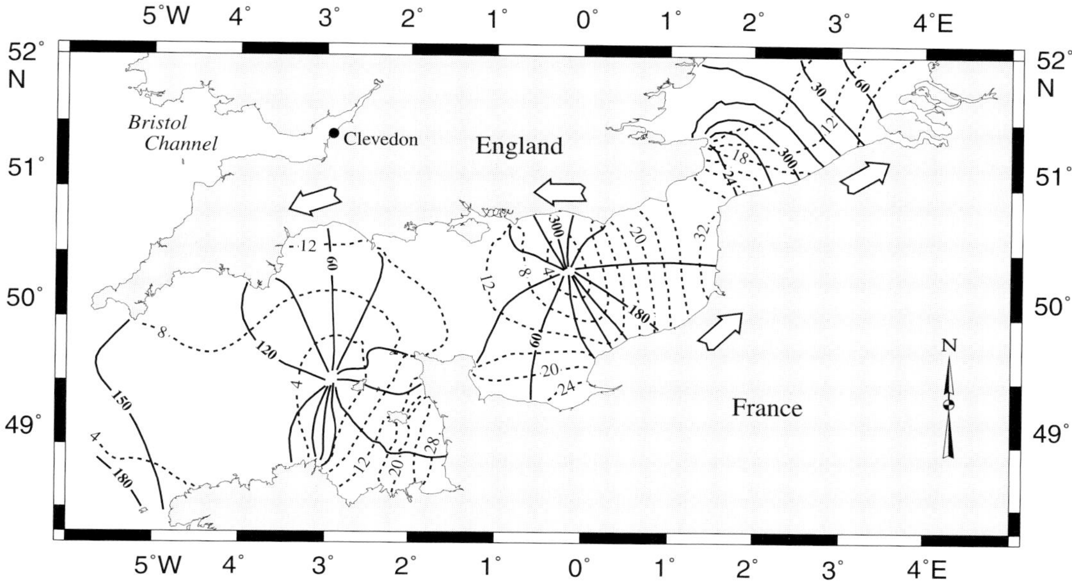

Figure 5.6. A tidal chart of the $\mathbf{M_4}$ shallow-water harmonic in the English Channel. Note the coherent pattern around amphidromes in mid-Channel (based on information supplied by Roger Flather).

are not limited to Southampton Water. At Portland, 90 km to the west of Southampton, there are sometimes double low water levels at spring tides.

We can conclude this discussion by returning to look at the separate harmonic constituents. In some cases it is possible to draw co-tidal and co-amplitude charts for the shallow-water constituents (Figure 5.6). For the sea around Britain the $\mathbf{M_4}$ amplitude is comparable to the amplitude of the astronomical diurnal constituents, in the range 0.05–0.15 m. The amplitude of $\mathbf{M_6}$ is smaller, but is particularly large in the vicinity of Southampton Water and the Isle of Wight, perhaps due to local resonances, as discussed. Generally $\mathbf{M_4}$ is the most important higher harmonic. Harmonics higher than $\mathbf{M_4}$ are only significant in estuaries and restricted local areas. The importance of these distortions for tidal processes at the coast will be discussed in Chapter 9. Even where they are relatively small, their influence on even more non-linear geological and biological processes, often in regions of intense human activity, can be very significant.

5.4 Currents in channels

Strong tidal currents are often found in straits joining two sea areas where different tidal ranges and times (and phases) prevail at the two ends. These strong currents are driven by the pressure head generated by the differences in sea levels acting along the short distance of the strait or channel. This distinguishes them dynamically from the currents due to tidal wave propagation discussed in Chapter 4. Currents in channels are commonly called *hydraulic currents*. Slack water with zero currents in the channel occurs when the levels are equal, and this may be at times quite different from the times of high or low water at either end. These channels are often important for ships' passage, and so an understanding of their currents is essential.

Similar currents are also found through narrow or shallow entrances to bays, fjords and harbours. Some of the more spectacular examples of tidal currents in channels include flows through the Straits of Messina between Italy and the island of Sicily, flow between the Indonesian islands and flow in the East River, New York between Long Island Sound and New York Harbour. The reversing falls in St John River, New Brunswick, Canada, and the similar reversing falls, the Falls of Lora, at the entrance to Loch Etive, Scotland are due to oscillating tidal currents across the shallow entrance.

For these channels, which are short compared with the tidal wavelength, the steady hydraulic currents are essentially a balance between the pressure gradient due to the slope on the water surface that drives the flow and the bottom friction that opposes the flow. Using the quadratic friction law in Equation (5.1), for a channel of length L where the mass of water to be moved depends on the depth D:

$$u = \left\{ \frac{Dg}{C_{\mathrm{D}}L}(\zeta_1 - \zeta_2) \right\}^{\frac{1}{2}} \tag{5.6}$$

The values of ζ_1 and ζ_2, the levels at the two ends of the channel, may be expressed as the sum of separate sets of harmonic constituents, and so used as the basis for current predictions. Note that this expression does not depend on the water density. The most direct approach for using this relationship to predict tidal currents in a particular channel is to calibrate it by measuring flows over a few tidal cycles and plotting them as a function of the square root of the water level differences across it. This automatically allows for the many complicated physical processes such as boundary turbulence, channel meanderings and effects at the open ends of the channel.

Figure 5.7 shows the water levels observed by NOAA at two places around New York on 5 October 2002. Kings Point at the top of Long Island Sound has larger tides due to the quarter-wave resonance discussed in Chapter 4. The timing of the tidal phases there is very different from those of the smaller tides at The Battery at the southern end of Manhattan

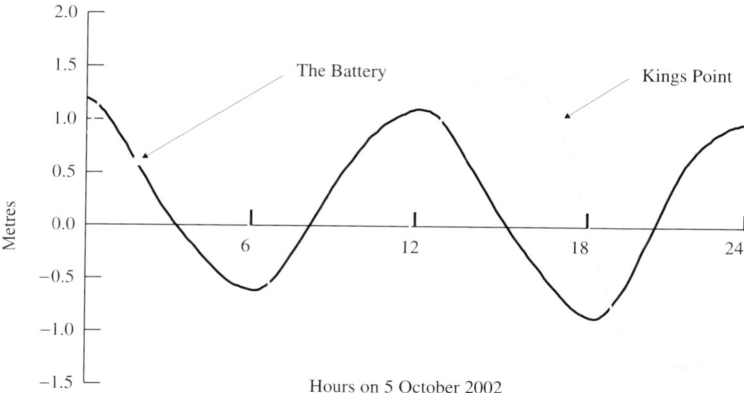

Figure 5.7. Sea levels observed by NOAA on 5 October 2002 at either end of East River, New York showing the strong surface slopes that drive the tidal currents in the channel.

Island, which are driven directly from the Atlantic Ocean. The level differences, which can reach 2 m between these places that are only 15 miles apart, produce the hydraulic gradient to drive the currents in the East River; the phases of these gradients are quite different from the phases of the levels. Elsewhere in New York Harbour, complicated tidal currents and phases are found in Hudson River and Harlem River.

For tidal hydraulic flow in straits discussed here to be important, the volume of water exchanged between the two sides must be too small to change the independent tidal wave dynamics of the separate sea level regimes at either end. The channel currents discussed in Section 9.1.1 are a slightly different form of this hydraulic process, where the levels at either end are not independent.

The very strong currents associated with channel flow sometimes give rise to strong eddies or vortices that occur at fixed times in the tidal cycle. Some, such as the Maelstrom in the channel between Moskenesoy and Mosken in northern Norway, and the Charybdis and Scilla whirlpools associated with the flow through the Straits of Messina are more celebrated in literature than they are in oceanography. There is a popular legend telling that the strait in which the Maelstrom develops is bottomless; however, detailed surveys show that in fact the water is relatively shallow, with a maximum depth of 36 m.

5.5 Tides in estuaries and rivers

Sea level records from shallow-water coastal locations, particularly estuaries, normally show that the interval from low to high water is shorter than the interval from high to low water: the rise time is more rapid than the fall, as shown in Figure 5.1. The tides at Southampton in Figure 5.3 are very unusual in showing a different pattern, but here we return to

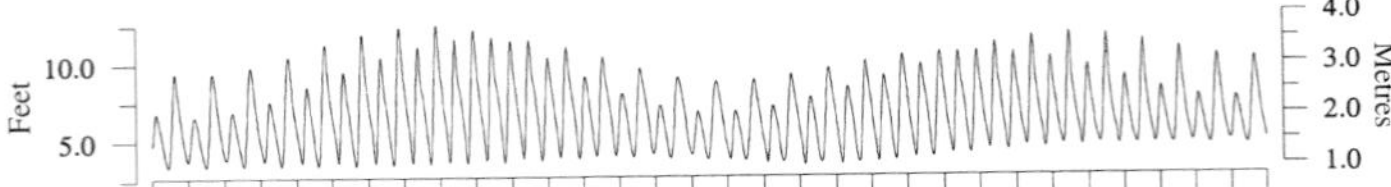

Figure 5.8. A typical month of tidal levels at Grondines in the upper reaches of the St Lawrence River, Canada. Low water levels vary much less than high water levels, and the lowest levels are found around neap tides (with acknowledgement to the Canadian Hydrographic Service).

consider the more normal situation. Offshore, the flood currents into an estuary or river are almost always stronger than the ebb currents. This asymmetry means that tidal currents tend to move sediment in preferred directions, as we shall discuss in more detail in Chapter 9. In terms of the circulation of water in estuaries, tides and the mixing they effect (which we shall also discuss later) cause density gradients that influence the circulation of water. Here we examine the distortions as tides travel into the higher reaches of rivers and the generation of the ultimate tidal distortion, the tidal bore.

5.5.1 Spring–neap effects

In the upper reaches of estuaries the tidal flow is strongly influenced by seaward flow in the rivers; these eventually exceed and reverse the tidal flow. Also, in very shallow waters frictional resistance becomes the dominant physical process over most of the tidal cycle with slower drainage at low water when the depth D is small, as shown by Equation (5.5.6). The time taken for the water to drain away after high water under gravity becomes longer than the tidal periods themselves, with the result that long-term pumping-up of mean water levels occurs over the period of spring tides. In the upper reaches of shallow rivers the lowest levels occur at or after neap tides, in contrast to normal coastal conditions where the lowest water levels occur at low water on a spring tide. Often the low water levels change very little over the spring–neap tidal cycle, compared with the much larger changes in high water levels.

A plot of a month of typical tidal levels at Grondines in the upper reaches of the St Lawrence River in eastern Canada shows this effect very clearly (Figure 5.8). The tides propagate up the St Lawrence River for more than 600 km, as far as Lake St Peter, over a period of 10 hours from Seven Islands. In the upper reaches of the river the high waters at different places have roughly the same level, but the low waters have a considerable slope downstream due to slower drainage. Above Quebec City, the low water level is higher on spring tides than at neaps because there is not enough time for the water to drain away against frictional forces before the next high water of the tidal wave arrives. In a tidal analysis this effect shows in the $\mathbf{MS_f}$ constituent (see Section 5.2). Similar effects are found in many other rivers, but the tidal influences are sometimes elusive at the upper limits as water levels are also very sensitive to the amount of water flowing down the river at any particular time.

5.5.2 Tidal bores

Many important towns and harbours are situated on estuaries or towards the highest navigable point of major rivers. The distortions of the tides that occur in these narrowing and shallowing regions can be both unusual and extreme, and are of considerable practical interest. The ultimate and most spectacular distortion of the tide is the generation of a tidal bore.

The gradual narrowing of an estuary tends to increase the tidal amplitudes. A progressive wave conveys energy at a rate proportional to the square of the amplitude and proportional to the width of the wave front. If energy is being transmitted at a constant rate to an area of bottom friction and energy loss, as in the upper reaches of an estuary, then as the estuary narrows a reduction of the width of the wave front must result in an increase of the wave amplitude:

$$H_0^2 \propto \frac{1}{\text{width of wavefront}}$$

Of course, this relationship will not hold strictly because although we assume that energy dissipation is confined to the upper reaches, in practice it will occur to some extent along the whole length of an estuary.

In the upper reaches of a narrowing estuary the wave crest may catch up with the trough, forming a tidal bore, analogous to a wave breaking on a beach. Many rivers have small tidal bores on certain tides, but large tidal bores are seen in only a few rivers. The basic requirement for a large bore is a large tidal range, but this condition alone is not sufficient. The main characteristic of a bore is the very rapid rise in water level as its front advances past an observer; from the riverbank it often appears as a breaking wall of water a metre or so high which advances upstream at speeds up to 5 $\mathrm{m\,s^{-1}}$ or greater. As energy is dissipated, the speed of the bore is reduced until, at an advanced stage of decay, its progress up the river may eventually be stopped and even reversed back downstream by the river flow.

Bores are known by many other names in the regions where they occur. The bore in the River Amazon is called the pororoca; in the River Seine, mascaret; and in the River Trent, aegir or eagre. The famous bore on the Qiantang River in China (Figure 5.9) is reported to have a height in excess of 2 m and to advance at speeds in excess of 6 $\mathrm{m\,s^{-1}}$; lives are still lost as spectators approach too close. Other impressive but less spectacular bores are found in the Petitcodiac and Salmon rivers which

Figure 5.9. The spectacular tidal bore on the Qiantang River, China, which can exceed 2 m in height and travel at speeds greater than 6 m s^{-1} (provided by Professor Lin Bingyao, Zhejiang Institute of Hydraulics and Estuaries).

flow into the Bay of Fundy, in the Hooghly River in India, the River Indus in Pakistan, the Colorado River in Mexico, the Turnagain Arm, Cook Inlet, Alaska and the Victoria River, Australia.

Bores are essentially due to instabilities in the hydraulics of tidal flow, and are therefore very sensitive to changes in river morphology. Increasing the river depth by dredging and speeding the river flow by building embankments both reduce the chances of a bore developing. Many bores have been reduced in amplitude by river engineering works; in some rivers there have been deliberate attempts to reduce bores because of the damage they can cause to riverbanks. The pattern of silting and dredging at the entrance to the River Seine has reduced the size of the mascaret. Similarly, the burro, the bore in the Colorado River, has been progressively reduced by siltation and land drainage schemes.

5.6 Tidal energy: turbulence and dissipation

5.6.1 Bottom friction

The drag of the sea bed on tidal currents opposes the flow (Section 5.1.1) and removes energy from the tidal wave. Here we look in more detail at the local mechanisms of energy loss, discussed in a global context in Section 4.6. The energy lost by the tides through bottom friction in shallow water is converted to turbulence, and eventually generates a small amount of heat. It is also an important agent in the shaping of the coast and the sea bed, and helps to mix the fresh water that flows into the coastal seas through the rivers. Equation (5.5.1) shows that the friction of the sea bed on a tidal current is related to the square of the current speed. In physics, the rate of energy loss is equal to the force multiplied by the distance moved by the force. We can translate this as:

$$\text{Rate of energy loss} = \text{Current speed} \times \text{bottom stress}$$

Consider a harmonically varying $\mathbf{M_2}$ tidal current in a channel $u = U_{M_2} \cos 2\omega_1 t$. Using Equation (5.1) for the bottom stress due to the frictional resistance to the current:

$$\text{Rate of energy loss} = C_D \rho U_{M_2}^3 \cos^3 2\omega_1 t$$

taking always the positive value. Averaged over a complete cycle of $\mathbf{M_2}$, the mathematics gives:

$$\text{Rate of energy loss} = \frac{4}{3\pi} C_D \rho U_{M_2}^3 \text{ per unit area} \qquad (5.7)$$

This shows that even a small increase in the speed of a tidal current substantially increases the energy lost. This 'cubic law' of energy dissipation implies that tidal energy loss is a strictly localised phenomenon concentrated in a few areas of shallow seas around the world where the currents are strongest (see Section 4.6).

Because of the cubic law, the energy lost by combinations of harmonic constituents is substantially greater than the energy that would be lost if they were considered separately. If we consider the average energy losses through a spring–neap tidal cycle, it can be shown that for the $\mathbf{M_2}$ and $\mathbf{S_2}$ harmonic tidal constituents, over a complete spring cycle:

$$\text{Average rate of energy loss} = U_{M_2}^3 \left(1 + \tfrac{9}{4}\gamma^2 + \cdots\right)$$

The factor $\gamma = U_{S_2}/U_{M_2}$; for an Equilibrium Tide the ratio $\gamma = 0.46$, which implies that the average energy dissipated over a spring–neap cycle is 1.48 times that dissipated by $\mathbf{M_2}$ alone. Around the northwest European shelf, where $\gamma = 0.33$ the average dissipation over a spring–neap cycle is 1.25 times the $\mathbf{M_2}$ dissipation. In the Equilibrium case the rate of tidal energy dissipation at springs is twenty times the rate of dissipation at neaps; in the case of the European shelf the spring dissipation exceeds the neap dissipation by a factor of eight. In the Minas Basin, Bay of Fundy, where $\mathbf{S_2}$ is relatively small, the energy dissipation on spring tides is only 2.5 times greater than that at neaps.

5.6.2 Moving amphidromes

The changing rates of energy losses in shallow seas result in a systematic adjustment of the tidal patterns during the spring–neap cycle. We have already considered the case of a standing wave on a rotating earth, which produces an amphidromic system (Section 4.2.3) as illustrated in

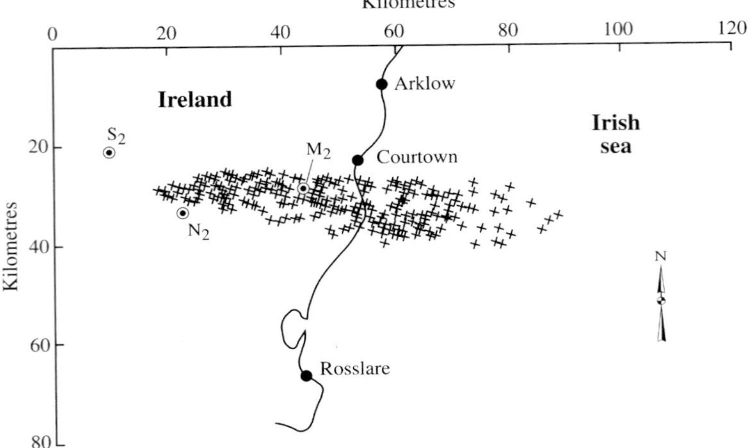

Figure 5.10. Detailed daily movements of the semidiurnal amphidrome at the southern entrance to the Irish Sea. The amphidrome is inland, virtual or degenerate, at spring tides. It is real for small neap tides. The positions of the individual $\mathbf{M_2}$, $\mathbf{S_2}$ and $\mathbf{N_2}$ amphidromes are also shown.

Figure 4.6. If the wave is not totally reflected at the head of the channel, the returning wave will have a smaller amplitude. As a result, in the northern hemisphere the amphidromic point will be displaced from the centre line of the channel towards the left of the direction of the ingoing wave (Figure 4.6b). A very good example of this amphidrome displacement can be seen in the distribution of the $\mathbf{M_2}$ amphidromes in the North Sea (Figure 4.10).

Normally, the positions of the amphidromes are plotted for individual harmonic tidal constituents as we have done for $\mathbf{M_2}$ and $\mathbf{K_1}$ and in Figure 4.7 and Figure 4.8. However, the amphidrome may also be considered as a moving, time-dependent position of zero tidal range. Its position can be fixed on a daily basis for each tidal *species* by plotting the $\mathbf{D_2}$ harmonic, which represents the total semidiurnal tide, as discussed in Section 5.3. When the daily positions of the semidiurnal $\mathbf{D_2}$ amphidrome at the southern entrance to the Irish Sea are plotted (Figure 5.10), they are found to move regularly back and forth within a narrow area: the range of the east–west movement exceeds 70 km, whereas the north–south movement is only 14 km. Within this narrow area the $\mathbf{D_2}$ amphidrome moves systematically, with a period of 14.8 days.

The maximum displacement from the Irish Sea centre coincides with spring tidal ranges, whereas the minimum displacement occurs at neaps. For neap tides the $\mathbf{D_2}$ amphidrome is real, with zero tidal range occurring at a point within the Irish Sea. At spring tides the amphidrome is inland as a virtual or degenerate point. This explains why the Courtown tides shown in Figure 2.1 are so unusual: as the amphidrome passes near Courtown in its spring–neap excursions, the semidiurnal tides almost disappear.

The reasons for this pattern of amphidrome displacement are related to the cubic law of energy dissipation, which means that much more energy is absorbed from the Kelvin wave at spring tides than is absorbed at neap tides. As a result, the reflected wave is relatively much weaker at spring tides and so the amphidrome displacement from the centre is greater. Similar spring–neap movements have been plotted for other amphidromes.

5.6.3 Tidal turbulence

Tidal currents in shallow regions lead to strong turbulence and stirring of the waters. A tidal current of 1 m s^{-1}, which is quite common, has a drag equivalent to that of a wind of near hurricane force (33 m s^{-1}). The drag takes energy from the tidal currents and converts some of it into turbulence; the turbulence decays slowly as energy moves into smaller and smaller eddies, and eventually produces a small heating of the water. There is often a competition between the rates at which surface heating or influx of low-salinity water tends to stratify the water into layers and the ability of the turbulence to erode and prevent stratification. In Section 4.5 we saw how internal tides can cause mixing at the boundary of the continental shelf and the deep ocean. In shallow water there are two further cases of special interest: seasonal stratification caused by summer heating of the surface layer, and stratification caused where low-salinity water from rivers and estuaries forms a layer above the seawater in the coastal zone.

In a region where the surface heating is approximately constant over a large area, the water column becomes completely mixed when the factor (depth/(current speed)^3) exceeds some critical value. The cube of the current speed is involved because of the processes described in Section 5.6.1. Where that critical value is equalled there is a boundary or *front* between mixed and stratified water. The critical value cannot be determined from theory alone because it depends on the efficiency of the mixing, i.e. the fraction of the turbulent energy available to overcome stratification. Typically this is less than one per cent. The frontal boundary is usually well defined with strong horizontal temperature gradients (often 10°C km^{-1}) between the warmer surface water in the stratified region and the cooler mixed surface water. Tidal fronts are easily identified on thermal satellite images. They are often areas of high biological activity and potential fish catches. In winter these fronts disappear as the surface waters become colder than the main water mass, so that mixing through the depth takes place by normal convection.

The other type of front affected by tidal mixing is found near the coast. In some coastal areas, strongly influenced by freshwater flows, stratification depends on the river discharge rates and varies as the tidal currents change over a spring–neap cycle. These regions have been called ROFIs, Regions Of Freshwater Influence. Unlike seasonal

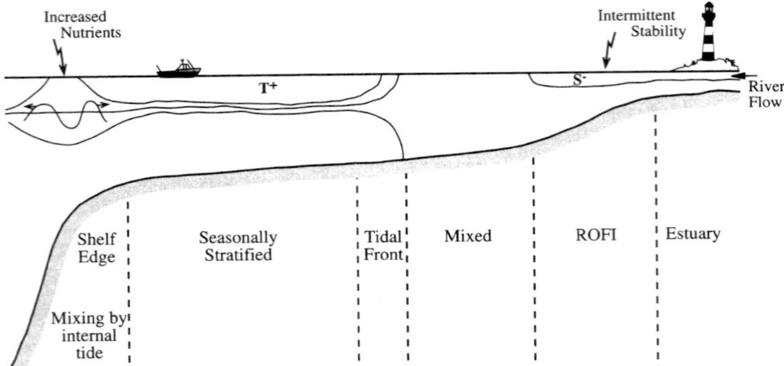

Figure 5.11. Locations of tidal mixing and fronts for tidal progression through a typical ocean–shelf–estuary system. Internal tides cause mixing at the shelf edge. Near the shore where freshwater flows in from the river, there is a Region of Fresh Water Influence (ROFI) (with acknowledgement to John Simpson).

Figure 5.12. The Annapolis Tidal Power Station, Nova Scotia, Canada. This has a working capacity of 20 MW, enough to provide power for a small town.

offshore fronts, the flux of fresh less dense water to sustain ROFIs is localised to rivers and estuaries, but the effects can spread extensively along and away from the coast. ROFIs are especially sensitive to the changes in tidal turbulence between spring and neap tides: there may be complete vertical mixing on spring tides and strong stratification at neaps. Once the sea is well mixed, it appears that stratification may only re-appear when turbulence falls below a critical level; stratification may not develop on all neap tides, but may only occur when they are especially weak, for example when the moon's apogee coincides with lunar quadrature.

Figure 5.11 shows a simple schematic of the different mixing regimes as a tidal wave moves from the deep ocean to an estuary. The first area of mixing is at the shelf break due to internal tides. On the shelf there is stratification until the tidal front, beyond which the currents are strong enough to mix the water completely. Finally, there is again stratification near the coast where freshwater fronts may form.

A tidal power scheme

Tidal energy schemes have been used for centuries. The remains of several ponds that were dammed and used to drive corn mills can still be seen along the coast of New England in the United States and around the coast of southern England.

The theoretically available energy in a tidal basin, filled at high tide and discharging to the sea at low tide, is:

$$2S\rho g R^2 \tag{5.8}$$

where S is the surface area of the basin, ρ is the water density and g is gravitational acceleration. R is the level difference between the sea and the tidally filled basin (twice the tidal amplitude). The ideal tidal site is one where a large surface area can be enclosed by building a barrier at relatively low cost across a narrow entrance. Because the energy depends on R^2, having a large tidal range is critical. If the turbines can be operated in a reverse sense as the empty basin is re-filled at high tide, the available energy is doubled.

Today, schemes that use modern technology for generating electricity from the tides have been operating for many years in three places. The largest of these at La Rance, near St Malo on the northern coast of France, with a capacity of 240 MW, has turbines that work as water flows in and out of the storage reservoir. A smaller scheme at Annapolis, on the Bay of Fundy coast of Nova Scotia (Figure 5.12), has a capacity of 20 MW. Completed in 1984, the scheme was built as a pilot to harness the outgoing tide with large turbines. A much more ambitious scheme to generate 200 times more energy from the tides by building a barrage across the Minas Basin is not now considered economically or environmentally viable. The third tidal power scheme, with a capacity of 0.5 MW, is located at Guba Kislaya, on the White Sea Arctic coast of Russia.

Engineers are also investigating the technical challenges of installing turbines in offshore tidal currents. The principle is similar to generating energy by wind turbines, but because the water density is much greater than that of air, the turbines can have a much smaller diameter. Tidal currents are also much more predictable than winds, and in a semidiurnal tidal regime reach maximum speeds four times a day. Also, submerged turbines are invisible from the shore and so are environmentally attractive. Designers favour speeds of between 2 and 3 m s^{-1}, as slower currents are uneconomical and stronger currents can damage the structures. There are many sites around the world where the tidal currents are suitable for power generating schemes, but the technical challenges are severe and developments are at an early stage.

Further reading

The complicated behaviour of tides in the coastal zone is mainly described in research papers, relating to a particular process or region, or both. Several of the reading recommendations at the end of Chapter 4 also apply here. Simpson (1998) gives a valuable synthesis of these. Some of the basic theory of wave distortion is covered in Lamb (1932), which remains a classic text. The tides and tidal currents of New York Harbour are described comprehensively in both Marmer (1926) and Redfield (1980), although their methods of analysis are more complicated than those used today. For readers interested in the problems of analysing and predicting shallow-water tides the chapter by Parker, Davies and Xing in Mooers (1999) is a useful introduction. Friedrichs and Aubrey (1988) review tidal distortions in estuaries. The movement of the Irish Sea amphidrome is described in detail in Pugh (1981).

Questions

5.1 Using Equation (5.5.3), calculate the amplitude of $\mathbf{M_4}$ for an $\mathbf{M_2}$ tidal wave of 6 m amplitude, travelling in a channel of 20 m depth, 20 km into the channel. Hint: the wavelength is given in Table 4.1.

5.2 From the example in Table 5.1, calculate the speeds in degrees per hour of the shallow-water harmonic constituents $\mathbf{MSN_2}$ and $\mathbf{S_4}$.

5.3 Estimate the amplitudes of the fourth-diurnal ($\mathbf{D_4}$) and sixth-diurnal ($\mathbf{D_6}$) shallow-water tides at Southampton from Figure 5.5, for an average $\mathbf{M_2}$ tide. Would the amplitudes of either, acting alone, be sufficient to cause a double high water? Assume $H_{\mathrm{M_2}} = 1.5$ m.

5.4 From Equation (5.6) and Figure 5.7, estimate the currents in New York East River at noon, assuming an effective distance of 25 km between The Battery and Kings Point. Why would the currents measured at a particular point be different from this? Assume a water depth of 30 m and a drag coefficient $C_{\mathrm{D}} = 0.0025$.

5.5 Use Equation (5.7) to calculate the ratio between the tidal energy available at a tidal power station at spring tides and that at neap tides. Assume $H_{\mathrm{S_2}} = 0.33 H_{\mathrm{M_2}}$.

Chapter 6
Weather and other effects

Even the most carefully prepared tidal predictions of future sea levels differ from those actually observed, because of weather effects and other factors. The relative importance of the tidal and the non-tidal components depends on the time of year, the severity of the weather and local water depths or bathymetry. Meteorological disturbances are usually greatest in winter, and have greatest effect where they act on shallow seas. The total level can give rise to serious coastal flooding when severe storms acting on an area of shallow water produce high levels that coincide with high water on spring tides. Where the surrounding land is both low lying and densely populated the inundations can result in human disasters of the greatest magnitude. Table 6.1 lists examples of such disasters. Some places, such as the Bay of Fundy, need only a moderate surge on top of a big tide for serious flooding. In other places very large surges can cause devastation on coasts that have only small tides, e.g. St Petersburg in the Baltic Sea. Even shallow lakes can cause serious flooding if acted on by extreme winds, as did Lake Okeechobee, Florida, USA, in 1928.

In this chapter we look mainly at the weather effects, air pressure and winds that continuously change the sea levels from those predicted using the tidal methods in Chapter 3. Total extreme levels due to combinations of big tides and extreme weather, and their impacts are considered in Chapter 8. At the end of this chapter we also look at two other factors that may affect sea levels. These are local resonant oscillations, called *seiches*, and waves caused by sub-marine seismic events, called *tsunamis* (see Figure 1.3). Both seiches and tsunamis are also strongly influenced by local water depths and the shapes of the coastlines on which they impact.

Table 6.1. *Estimated impacts of some historical storm surge events.*

Date	Region	Maximum surge level (m)	Lives lost	
November 1218	Zuider Zee	?	100 000	
September 1775	Newfoundland, Canada	?	4000	
1864, 1876	Bangladesh	?	250 000	
October 1869	Bay of Fundy, Canada	2.0	100	Saxby Gale
September 1900	Galveston, Texas	4.5	6000	
September 1928	Lake Okeechobee, Florida	2.5	1800	
September 1938	New England	3.5	600	Long Island Express
February 1953	Southern North Sea	3.0	2000	
March 1962	Atlantic Coast, USA	2.0	32	Ash Wednesday Storm
August 1969	Mississippi, USA	7.0	256	Hurricane Camile
November 1970	Bangladesh	9.0	250 000	
November 1978	Sri Lanka/Tamilnadu	4.0	373	
November 1988	Bangladesh	4.4	5708	
April 1991	Chittagong, Bangladesh	4–8	150 000	
August 1992	Miami, Florida	5.1	50	Hurricane Andrew
October 1999	Orissa, India	7–8	10 000	

6.1 Background

The surge is the difference between the observed and the predicted sea levels at any time, $S(t)$ as defined in Equation (3.1). Mean sea level is a separate component of the total observed sea level, and not part of the defined surge discussed here. The term *storm surge* is normally reserved for the excess sea levels generated by a severe storm. A more general term is *non-tidal residual*, which is sometimes also called the non-tidal component, or the meteorological residual. Hydrodynamically, the term *surge* implies a sudden movement of water that is quickly generated but which is soon over. Alternative popular descriptions for severe flooding events include freak tide, storm tide and tidal wave, none of which is valid in exact scientific usage.

If the weather effects increase sea levels there is a ‘positive’ surge; lowered levels are termed ‘negative’ surges. Positive surges can add to high tides and waves to cause coastal flooding. Negative surges reduce water depth and are a potential shipping hazard. The strong currents that can be associated with surges may be important agents for erosion and geological change.

There are two ways in which the weather affects sea levels. Firstly, changes in atmospheric pressure produce changes in the pressure forces acting vertically on the sea surface, which are felt immediately at all depths. Secondly, forces due to wind drag at the surface are generated at and parallel to the sea surface; the extent to which they are felt at depths below the surface is determined by the length of time for which they act and by the density stratification of the water column, which controls the downward transfer of momentum. Usually, because they are linked in any particular storm, the effects of winds and air pressures on the observed sea levels cannot be separately identified, although it is often convenient to try to do so.

The effects on sea level of tropical storms and of extra-tropical storms are different, and so a clear distinction is usually made between them, as we have done in Figure 1.3. The length and space scales are very different.

Tropical meteorological storms are small and very intense. They are generated at sea during particular seasons, from where they move in a relatively unpredictable way until they meet the coast. Here they produce exceptionally high flood levels within a confined coastal region of perhaps tens of kilometres. Tropical storms are known variously as hurricanes (USA), cyclones (India), typhoons (Japan), willi-willies (Australia) and baguios (Philippines).

Extra-tropical storms extend over hundreds of kilometres around the central region of low atmospheric pressure and are usually relatively slow moving. They affect large areas of coasts over periods that can extend to several days. Along the east coast of the USA extra-tropical storms are generally more important to the north of Cape Hatteras, whereas tropical hurricane storms are most important to the south.

6.2 Some statistics of meteorological residuals

We have already discussed the way in which non-tidal residuals are defined in Section 3.2. If a time series $S(t)$ of the hourly residuals is computed by subtracting the predicted tide and the mean sea level from the observed levels, several useful statistics may be derived. One of these, the standard deviation, is shown for some coastal sites in Table 6.2. The standard deviation of $S(t)$ (see Section 3.2) from the mean value of zero is a general measure of the size of the weather effects. It varies from a few centimetres at tropical ocean islands to tens of centimetres in areas of extensive shallow water subjected to stormy weather. Table 6.2 shows that Honolulu has a very low standard deviation (0.06 m) whereas Buenos Aires has a high value (0.49 m). These figures may be used to derive confidence limits and to answer the question: how good are tidal predictions? Of course they are not 'errors' in the tidal predictions in the sense that they could be avoided by using better prediction methods.

Table 6.2. *Standard deviations of non-tidal meteorological residuals at some representative sites.*

Location	Standard deviation (cm)	
Honolulu	6	Ocean island surrounded by deep water
Mombasa	4	East African estuary near deep water
Buenos Aires	49	Extensive estuary of River Plate, Argentina
Newlyn	15	Shallow site on northwest European shelf
Southampton	20	Shallow site with strong non-linear interactions
Courtown	15	Shallow site near Irish Sea amphidrome

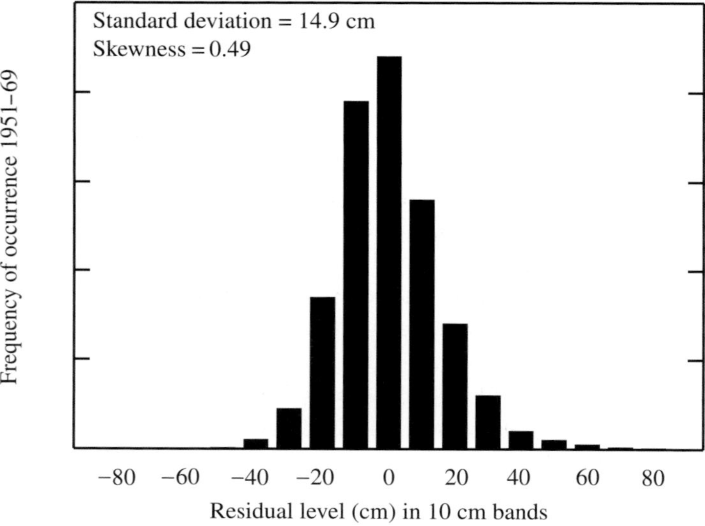

Figure 6.1. A histogram of Newlyn residuals (observed levels minus predicted tide) over a 19-year period, in 0.1 m intervals. Residual distributions have extended tails for both positive and negative events and are usually skewed towards positive events.

The frequency distribution of residual levels at Newlyn is plotted in Figure 6.1. Newlyn, in the southwest of England, is surrounded by the shallow-water region of the Celtic Sea, and faces the open Atlantic Ocean. Although the spread of non-tidal residuals looks very like a bell-shaped normal statistical distribution, there are important differences between the symmetrical normal distribution and the observed frequency distribution. The observed distribution of residuals has extended tails for both positive and negative events: these tails include the major surge events. There is also a tendency for large positive residuals to occur more frequently than large negative residuals, giving a positive skew to the distribution. As a general rule for most British ports, positive surges in excess of five standard deviations occur somewhat less frequently than

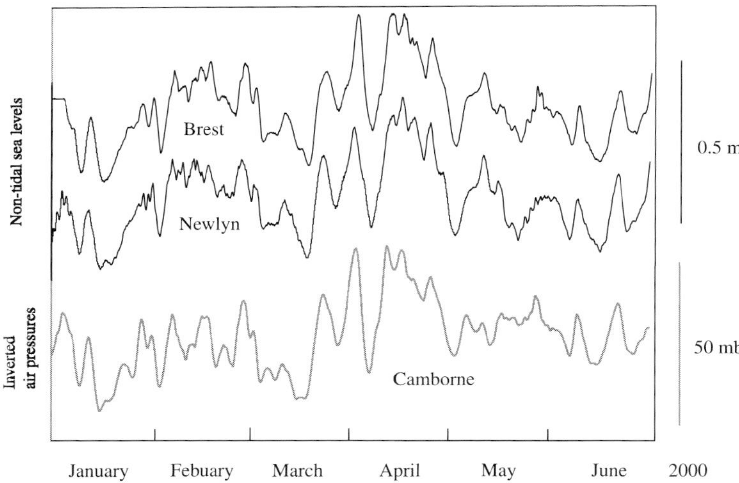

Figure 6.2. Non-tidal variations of the sea level around the English Channel at Newlyn and Brest, compared with air pressures at Camborne near Newlyn. The air pressures are plotted with the negative values at the top, clearly showing the inverted barometer effect.

once a year, whereas negative surges in excess of four standard deviations occur on average less than once every two years.

The weather effects on sea level are normally coherent over a wide area for storms in the middle and high latitudes. Figure 6.2 shows the subtidal (greater than a day) variations for Newlyn and Brest. The sea level changes are highly coherent with each other and also with the inverse of the low-frequency air pressure variations, plotted at the bottom of the diagram.

Casual inspection of Figure 6.2 suggests that the surge activity in the area of Newlyn and Brest has a preferred period of a few days, which might be attributable either to a dominant period in the atmospheric forcing or to a natural response of the shelf seas. However, this impression is misleading. A detailed analysis of Newlyn levels over 61 years (Figure 6.3), where the energy at each period is plotted as a spectrum, shows that no such peak exists. The smooth spectrum shows that there are non-tidal variations in sea level at all periods, with a gradual increase of energy as periods of 10 days and longer are approached.

The strong seasonal cycle in the weather is also present in the non-tidal residuals, but the seasonal changes do not always produce the biggest effects in the winter months. Figure 6.4 illustrates the statistics of residuals in two different ways. At Newlyn the most severe events tend to occur during the period November to April, while May to September are quiet months. However, severe storms can also occur during the late summer months, as shown by the well-known Fastnet storm, southwest of Britain, early in August 1979, when several racing yachts were lost. At Buenos Aires the most probable months for positive surges (defined as

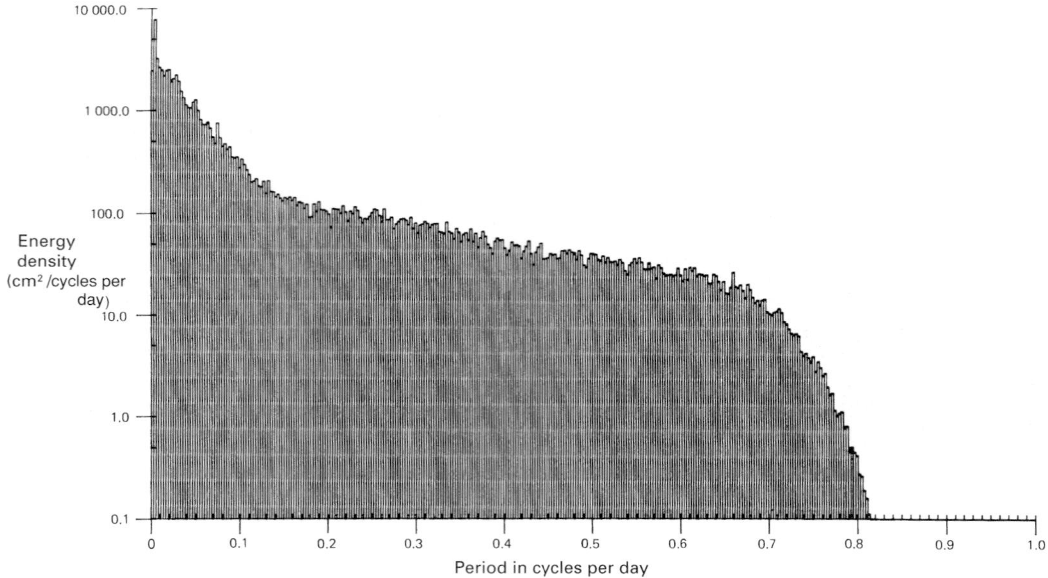

Figure 6.3. A detailed spectrum over 61 years of Newlyn sea-level changes at low frequencies. There is no evidence for resonant peaks. Instead, there is energy at all frequencies.

greater than 0.8 m and persisting for longer than 6 hours) are December and January.

The statistical description of sea level variations due to tropical storms is much more difficult than at higher latitudes because the events are so local. There are too few storm flooding events for them to be properly observed in a period of measurements made at a single site. Wherever an extreme flooding event has occurred it is afterwards possible to estimate the maximum levels from the damage caused; unfortunately, these observations are also too scattered, too rare and too unreliable to form the basis for estimating the probability of future floods. Alternative methods for estimating probabilities are discussed in the box (Flood warning systems) at the end of this chapter.

6.3 Responses to atmospheric pressure

In Figure 6.2 there is a clear link between the sea levels and the atmospheric pressure. If the level in a mercury barometer increases, the sea levels are depressed and vice versa. This response of sea level is called the *inverted barometer effect*. This inverse relationship between sea level and atmospheric pressure can be modelled theoretically. Suppose that the sea has reached an equilibrium condition in response to an applied

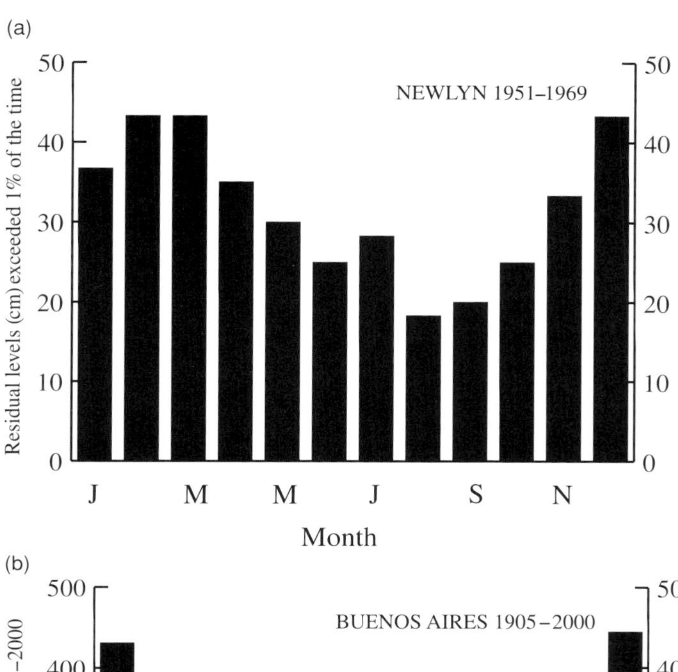

Figure 6.4. Two ways of showing the seasonal variations of weather effects on sea levels. (a) Seasonal changes in the value exceeded 1 per cent on the time at Newlyn. For a normal distribution this is equivalent to three standard deviations. Maximum weather effects occur between November and April. (b) The cumulative number of positive surges at Buenos Aires, Argentina (defined as greater than 0.8 m amplitude and longer than 6 hours duration). These are most common in the southern hemisphere summer between December and March (based on data supplied by Servicio Hidrografia Naval, Republica Argentina).

atmospheric pressure field so that there are no currents. Then for a horizontal level at depth in the sea the pressures will be equal to:

$$P_A + \rho g \zeta = \text{constant} \tag{6.1}$$

where P_A is the atmospheric pressure, ρ is the water density, g is gravitational acceleration and ζ is the sea level; this applies at all places, at the chosen horizontal level in the sea.

For local variations of atmospheric pressure ΔP_A about the mean atmospheric pressure over the oceans, the level of the sea surface will change relative to the mean sea level according to:

$$\Delta\zeta = -\frac{\Delta P_{\mathrm{A}}}{\rho g}$$

Taking values of seawater density $\rho = 1026\ \mathrm{kg\ m^{-3}}$ and $g = 9.80\ \mathrm{m\ s^{-2}}$:

$$\Delta\zeta = -0.993\ \Delta P_{\mathrm{A}}$$

where ζ is in centimetres and ΔP_{A} is in millibars. Therefore, an increase in atmospheric pressure of one millibar will produce a theoretical decrease in sea level of one centimetre. During a typical year extra-tropical atmospheric pressures may vary between values of 980 mb and 1030 mb. Compared with a Standard Atmosphere of 1013 mb, this implies a range of sea levels due to air pressure changes of between +0.33 m and −0.17 m.

This static inverted barometer effect can be enhanced for a moving atmospheric pressure system, especially if the system moves over the continental shelf at a speed ($c = (gD)^{\frac{1}{2}}$) close to the speed of a shallow-water wave (see Section 4.2.1), when a form of travelling resonance occurs.

Atmospheric pressures in tropical regions have a much smaller range, their chief characteristic being a 12-hour cycle with amplitudes around 1 mb and maximum pressures at 1000 and 2200 hours local time. The inverted barometer response of the sea levels to these pressure cycles produces a non-gravitational local tide with the same frequency as the gravitational solar semidiurnal tide, $\mathbf{S_2}$. This tide has a minimum at 1000 and 2200 hours local time.

It is rarely possible to identify separately the effects of atmospheric pressure disturbances as they are always associated with winds and the effects of wind stress, which we will discuss in the next section. The direct inverted barometer response is most likely to be observed alone when the sea is protected from wind stress by a covering of ice. Figure 6.5 re-plots the original results obtained by Sir James Clark Ross, who spent the 1848–49 winter in the Canadian Arctic whilst searching for the lost Franklin expedition. It shows a response very close to the theoretical static response (broken line) for an ice-covered sea, which Ross published as early evidence of the inverted barometer effect.

6.4 Responses to wind

The drag of the wind on the sea surface moves the water in ways that change sea levels. The biggest effects are observed when strong winds blow over shallow water. The theoretical response of sea levels to wind stress on a non-rotating earth is introduced first. However, because of

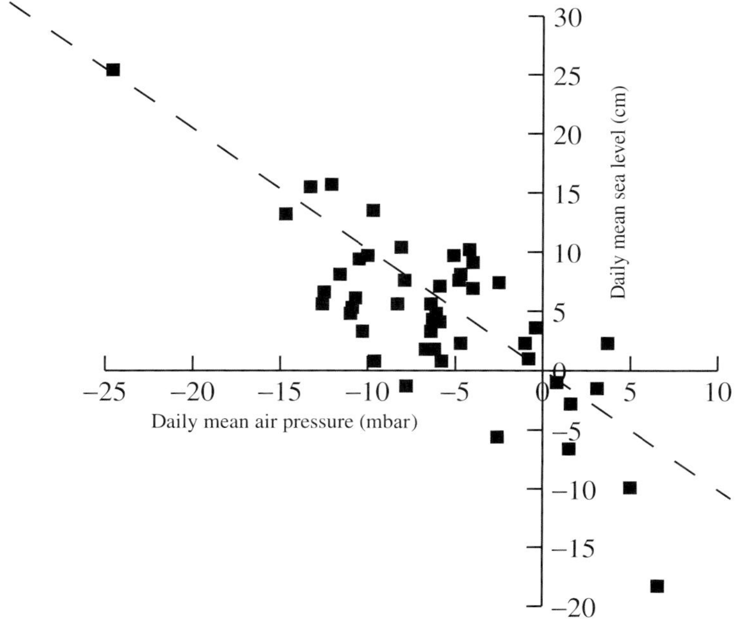

Figure 6.5. The data obtained by James Ross at Port Leopold in the Canadian Arctic (1 November to 17 December 1848). The daily mean air pressures and sea levels have been converted from the original units of inches of mercury and feet. Ross used these data in an early publication to demonstrate the inverted barometer effect.

the earth's rotation, some of the simple intuitive explanations turn out to be more complicated in practice.

6.4.1 Stress laws

When two layers of moving fluid are in contact, energy and momentum are transferred from the more rapidly moving layer to the slower layer. The physics of this transfer process is very complicated; however, basic functional relationships can be combined with empirical constants to give useful formulae for calculating some of the effects. As a general rule, the drag of the wind on the sea surface is given by:

$$\text{Drag} = C_D \rho_A W^2 \text{ per unit area} \tag{6.2}$$

where ρ_A is air density and C_D is a dimensionless drag coefficient. W, the wind speed, is usually measured 10 m above the sea surface. Not unexpectedly, this formula is similar to Equation (5.1) for current drag on the sea bed. The wind stress on the sea surface is in the direction of the wind and is proportional to the square of the wind speed. This means that although wind effects on sea level are small for normal winds, they can become very significant during big storms. Experiments have shown that the drag increases slightly more rapidly than the square of the wind speed. This increase with wind speed may be accounted for by an

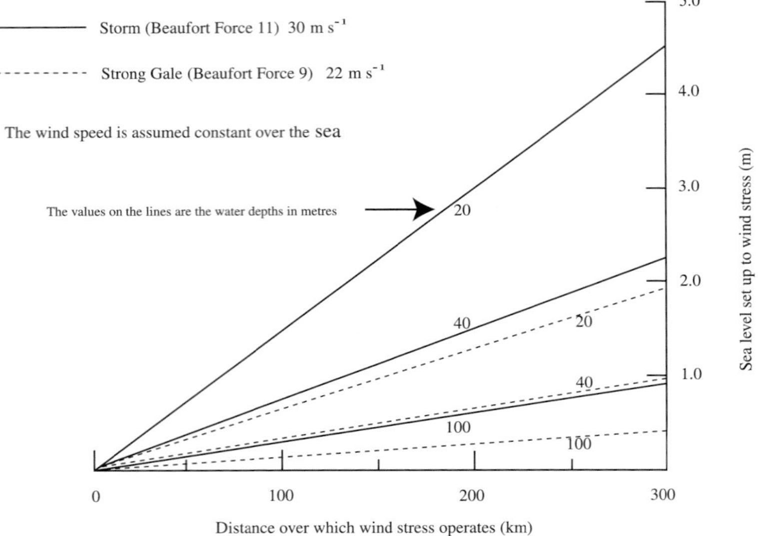

Figure 6.6. The theoretical relationship between wind speed, water depth and the slope on the sea surface. The winds are shown for Storm and Strong Gale for channels of depths 20 m, 40 m and 100 m on a non-rotating earth.

increase of C_D, and justified in terms of an increased surface roughness with increasing wave height. One form is ($10^3 C_D = 0.8 + 0.065W$), not very different for the range of C_D found for current drag on the sea bed.

6.4.2 Wind set-up

For a wind blowing along a narrow channel of constant depth the steady-state effect of wind drag on the slope of the sea surface is to pile water up at the downwind end and to produce higher sea levels there. This can be seen as a balance between the force of the wind stress and the pressure gradient that opposes it. In a steady state, neglecting effects of bottom friction, this slope on the sea surface can be written theoretically as:

$$\text{Slope} = \frac{\text{Increase in level}}{\text{Horizontal distance}} = \frac{C_D \rho_A W^2}{g\rho D} \tag{6.3}$$

Figure 6.6 shows the slope for some different wind strengths and water depths. The formula makes the important point that the effect of winds on sea levels increases inversely with the water depth D and so will be most important when the wind blows over extensive regions of shallow water. For a Strong Gale (Beaufort Force Nine, 22 m s^{-1}) blowing over 200 km of water which has a depth $D = 40$ m (approximately the dimensions of the southern North Sea) the increase in level would be 0.85 m. If the wind speed increased to Storm (Beaufort Force Eleven, 30 m s^{-1}) the level of increase would be 1.60 m. Of course the response of a sea to a wind will not be instantaneous; in practice the time taken to reach steady-state conditions will be comparable to the period of the earth's rotation and so rotational effects cannot be ignored.

6.4.3 Ekman transport

The Norwegian scientist and explorer, Fridtjof Nansen, when his ship *Fram* was stuck in the Arctic ice during 1893 to 1896, observed that the ice on the sea surface moved in a direction to the right of the wind and not in the direction of the wind itself. The Swedish oceanographer, Vagn Walfrid Ekman (1874–1954), explained this in terms of the hydrodynamic equations of water motion, including the earth's rotation. The net transport to the right of the wind stress in the northern hemisphere per metre of section perpendicular to the wind stress is given theoretically by:

$$\text{Transport} = F/f\rho \tag{6.4}$$

where F is the wind stress; f is the Coriolis parameter $2\omega_s \sin\phi$, where ϕ is the latitude and ω_s is the angular speed of the earth on its axis ($\omega_s = \omega_0 + \omega_3 = \omega_1 + \omega_2$); and ρ is the water density. Net transport is to the left in the southern hemisphere. This is called the *Ekman volume transport* per metre of cross-section, and is a function of the wind stress and the latitude. For a relatively strong wind speed of 15 m s^{-1}, at 45° latitude the wind-driven transport is 4.4 m^2 s^{-1}. For the more normal speed of 5 m s^{-1}, the transport is 0.29 m^2 s^{-1}.

For sea level studies the most important effects of Ekman transport are seen when the winds blow parallel to a coastline. The result is shown in Figure 6.7a. If the coast is to the right of the wind direction in the northern hemisphere, then sea levels are increased; if the coast is to the left then sea levels fall. The relationship is reversed in the southern hemisphere. The water level differences, caused by this initial Ekman transport, have their own influence on the water movements, so distorting the simple Ekman transport dynamics.

Again, from theoretical arguments, it can be shown that wind stress produces:

- a current parallel to the shore in the wind direction

$$u = \frac{Ft}{\rho D} \tag{6.5a}$$

- a slope on the sea surface at right angles to wind direction

$$= \frac{fu}{g} \tag{6.5b}$$

Equation (6.5a) shows the current parallel to the shore increasing steadily with time at a rate inversely proportional to the depth of water to be

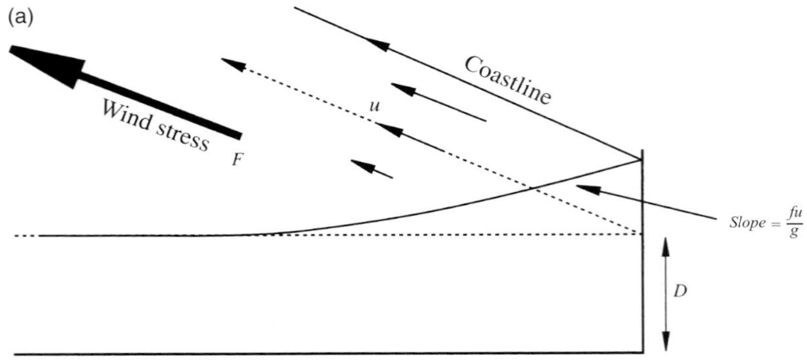

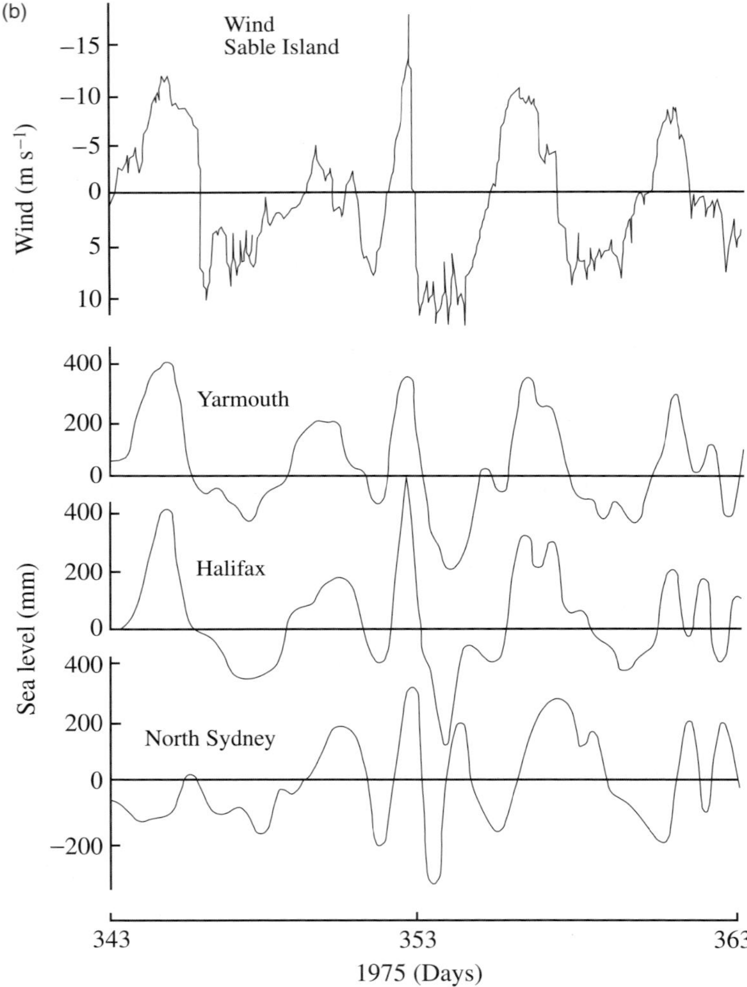

Figure 6.7. The effects of Ekman transport on sea levels. (a) The relationship between the transport of water and the wind direction in the northern hemisphere, and the surface slope. (b) Along-shore winds and residual sea levels along the Nova Scotia coast showing how winds from the northeast raise coastal sea levels on the right (from Sandstrom (1980), copyright, AGU).

moved, as would also be the case in the absence of the earth's rotation; in practice this current will eventually be limited by bottom friction. The sea surface slopes at right angles to the wind direction as shown by Equation (6.5b); the slope is upwards to the right in the northern hemisphere and to the left in the southern hemisphere (see Figure 6.7a). Equation (6.5b) implies a steady increase of coastal sea levels with time. Both currents and sea level gradients are greatest when the water is shallow.

The net result of the applied wind stress is a flow of water along the coast in the direction of the wind, as we would expect intuitively. However, we can now appreciate that this apparently simple direct relationship is really more complicated: it results from the condition that no water can flow across the coastline. This leads to a steady increase of coastal levels in response to Ekman transport and this increase in turn develops in dynamic equilibrium with the along-shore current. This sea level change is called the *locally* generated surge, to distinguish it from surges propagating freely as progressive waves that have travelled from external areas of generation.

The close correlation between along-shore winds and coastal sea levels is clearly shown in Figure 6.7b, which relates levels on the coast of Nova Scotia to the offshore winds. Winds blowing to the west, with the coast on the right, cause high coastal sea levels. Winds to the east give negative surge residuals.

6.5 Some regional examples of surges

So far we have considered the statistics of surge residuals and some basic ideas of meteorological forcing by air pressure and winds. It is now appropriate to consider some examples from a few well-studied regions, showing the physical processes that combine to give larger surges. As discussed, there is a natural distinction between the effects of storms at low latitudes (tropical surges) and those at higher latitudes (extra-tropical surges). The latter are slower to develop and more widely spread in their impact than the intense local impacts of tropical surges.

The North Sea

The North Sea, between Britain and northern Europe, has been described as a splendid sea for storm surges. It is open to the North Atlantic Ocean in the north so that extra-tropical storms, which often travel across this entrance from west to east, are able to set the water in motion with very little resistance from bottom friction. Surges are generated by winds acting over the shelf to the north and northwest of Scotland, and by pressure gradients travelling from the deep Atlantic to shallow shelf waters. When these water movements propagate into the North Sea they are affected by the earth's rotation and by the shallowing water as they approach the narrowing region to the south. These disturbances are sometimes called *external surges* to distinguish them from the movements and changes of level brought about by the wind acting on the sea surface within the North Sea (Section 6.4.3), which are called *internal surges*. The rate of progression of the external surge is controlled by the water depth according to $c = (gD)^{\frac{1}{2}}$. The time to travel between the Forth Estuary in Scotland and the Thames Estuary is about nine hours, which allows time for flood warnings to be issued along the coast of England and The Netherlands. London is protected by the Thames Barrier. The Oosterschelde Flood Barrier and the other defences of the Delta Works protect The Netherlands, much of which is below sea level and densely populated. Both depend on this natural surge progression to give time for the defences to be activated (see the box on Flood Warning Systems at the end of this chapter).

West coast of the British Isles

The dynamical characteristics of surges on the west coast of the British Isles are different from those in the North Sea. The sub-tidal oscillations shown in Figure 6.2 form a continuous weather-driven variation of the sea level in the English Channel: this is linked to the barometric pressure variations according to the inverted barometer relationship. Those variations that are not accounted for by an inverted barometer response can be correlated with the local winds. The most effective wind direction for producing large surges is from the south and southeast, across the mouth of the English Channel, in accordance with the Ekman transport of water to the right of the wind stress, and the corresponding build-up of coastal sea levels.

The second difference between west coast surges and those observed in the North Sea is related to the near-resonant response of the Irish Sea and Bristol Channel to tidal forcing from the Atlantic. It appears that these resonant modes are also responsive to meteorological forcing at similar frequencies, resulting in short-lived intense surges which are quickly generated and which decay during a single semidiurnal tidal cycle. A spectacular example of this type of surge occurred at Avonmouth in the upper reaches of the Bristol Channel in March 1947 when a surge of 3.54 m on a falling tide reversed the normal tidal fall of sea level for several hours.

Atlantic coast of North America

Around the Gulf of Mexico and along the Atlantic coast from Florida to Cape Hatteras the greatest risk of flooding comes from tropical storms – hurricanes – that originate in the tropical Atlantic Ocean from where they travel in a westerly direction until they reach the West Indies. Here many of them turn northwards towards the coast of the United States. Their greatest effects on sea level are confined to within a few tens of kilometres of the point at which they hit the coast (Figure 6.8). The Galveston 1900 surge (Table 6.1) is a spectacular example. In 1969, Hurricane Camile raised levels by up to 7 m on the Mississippi coast, and more recently Hurricane Andrew in 1992 raised sea levels locally in Miami by 5.1 m.

Further north, along the coast of the middle Atlantic States, the surges due to extra-tropical storms are dominant. The winds are less extreme, but the effects are more widely spread over hundreds of kilometres. The Ash Wednesday Storm of March 1962 caused flooding and coastal damage from North Carolina to New York. An intense depression developed over 5–7 March some 500 km east of Cape Hatteras. The strong winds from the northeast and east associated with this depression were experienced along the whole eastern seaboard; a combination of onshore wind set-up, Ekman transport towards the coast due to long-shore winds and spring tides resulted in some of the worst flooding ever recorded. The high surge levels persisted over five tides. Most of the direct destruction was due to severe beach erosion with breaking waves acting on top of the extreme sea levels. Hurricanes sometimes travel as far north as New England, as in the Long Island Express storm of 1938, but the areas most at risk are further south, where the danger of flooding is also greater because the coastal area is low lying.

Around the Canadian Maritime Provinces, severe flooding can easily be triggered by surges of only one or two metres on top of big spring tides. The notorious Saxby Gale flooding of 1869 occurred with a surge of only about 2 m, on an equinoctial spring tide. The close correlation between winds blowing parallel to the coast and coastal sea levels, a result of Ekman Transport, is clearly shown in Figure 6.7b, which relates sea levels on the coast of Nova Scotia to the offshore winds at Sable Island. Figure 6.9 shows a recent surge at Charlottetown, Prince Edward Island, due to an intense low-pressure system from the Atlantic.

Bay of Bengal

The coasts of India and Bangladesh that surround the Bay of Bengal are very vulnerable to severe flooding due to tropical cyclones. These cyclones usually originate in the southern parts of the Bay or in the Andaman Sea, further south, from where they move towards the west before curving to the north and northeast. In the north, where the shelf is 300 km wide, a severe cyclone in November 1970 hit the coast north of Chittagong and produced surge levels of more than 9 m. The fact that

(a)

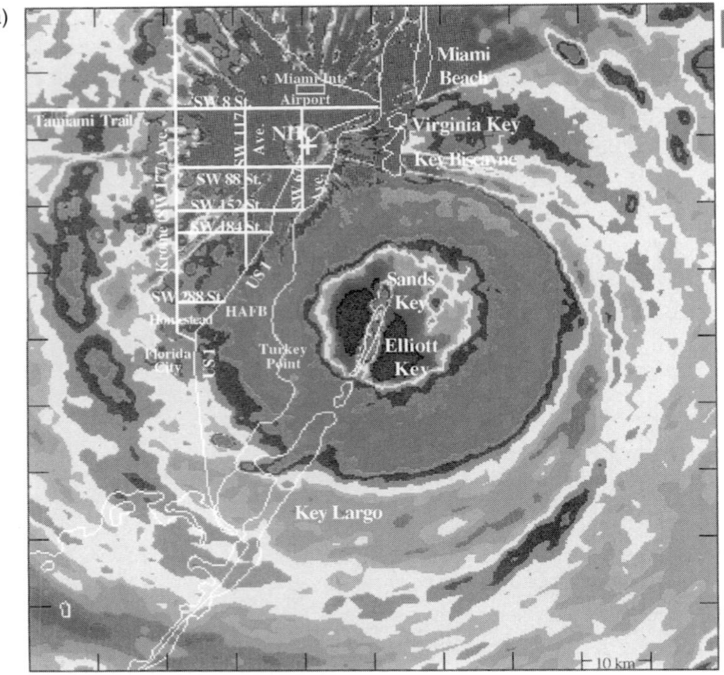

(b)

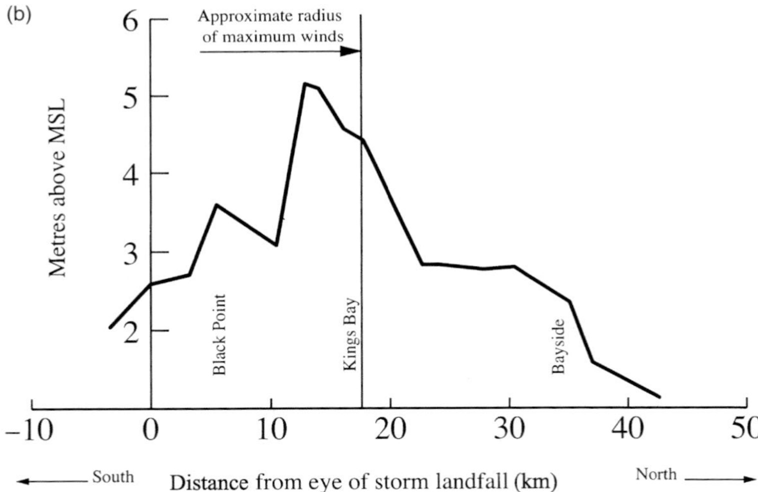

Figure 6.8. Details of the 1992 Hurricane Andrew approaching Miami where it caused severe damage (image and data supplied by NOAA/AOML). (a) A radar image of the storm as it approaches Miami. The map shows 100 km × 100 km. (b) The effects of the hurricane on sea levels along the coast. The maximum effects are concentrated over about 30 km.

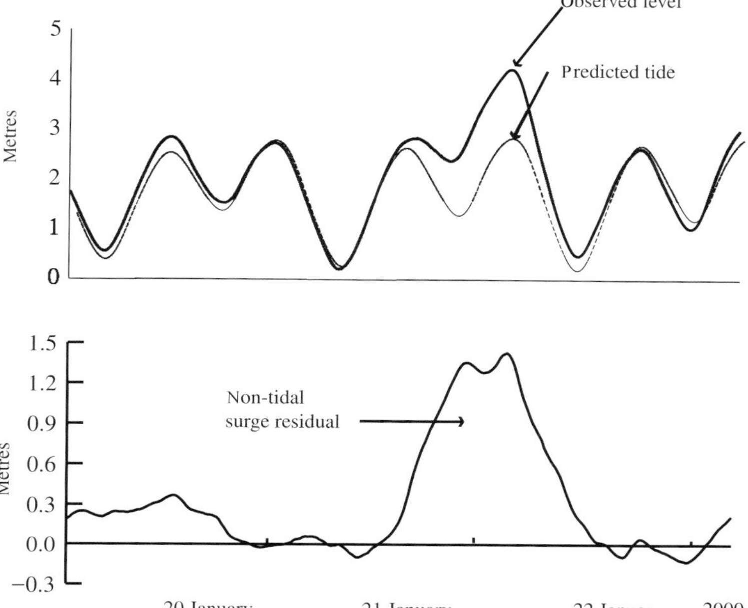

Figure 6.9. An example of an extra-tropical storm, from Charlottetown, Prince Edward Island, Canada in January 2000 (based on data supplied by the Canadian Hydrographic Service).

the coastline has a right-angle turn near Chittagong produces maximum surge levels, higher than those which the same storm would produce if moving perpendicular to a straight coast. Although the 1970 storm was one of the most exceptional in modern times, there have been many other recent examples, including the storm of April 1991. The nearby Indian coast of Orissa had a positive surge of more than 4.5 m in June 1982.

Japan

Several thousand people were killed in September 1959 when Typhoon Vera struck Japan, producing a peak surge of 3.6 m at Nagoya on the south coast of Honshu. The outer coast of Japan has a very narrow continental shelf, so that typhoons there have their greatest effect on individual bays. The wind effects are less important here than atmospheric pressure changes. If the direction and speed of a typhoon produces a resonant response of the natural period of the bay, then the effects can be particularly severe. Oscillations of the waters of a bay, called *seiches*, persist for several hours once they are set in motion. These oscillations, which are also excited by tsunamis, are discussed in Section 6.6.

The Adriatic

The increasing frequency with which the city of Venice, at the head of Adriatic Sea, is subject to flooding by high sea levels has attracted much attention in recent years (see Figure 8.8). These flooding events happen more often than in historical times because of a gradual increase of mean sea level, relative to the land, of between 3 and 5 mm per year (see Chapter 7). Storm surges in the Adriatic Sea are most effectively generated by pulses of strong winds from the southeast, directed along its length. These winds are associated with depressions that move eastwards from their region of development over the Ligurian Sea to the northwest of Italy. Observations of the surges produced at Venice show oscillations with periods near 22 hours that persist for many days. These oscillations, which are due to excitation of the fundamental longitudinal oscillation of the Adriatic Sea, have a maximum amplitude sometimes in excess of 1 m. One of the difficulties in forecasting these surges is the poor estimation of surface wind speeds over the Adriatic; these winds show considerable differences from one place to another due to the effects of the surrounding mountain ranges.

The Baltic Sea

The Baltic Sea has a very small tidal range due to its limited connection to the North Sea through the Kattegat and Skagerrak, but it is shallow and subjected to severe extra-tropical storms that generate large surges. One of these surges raised sea levels by 4 m at St Petersburg in 1924; earlier events occurred in 1777 and 1824. Defences for St Petersburg are now being built across the River Neva: eleven dams can be closed when flooding is expected, usually due to the approach of a depression across the Baltic Sea from the southeast. In the winter, when there is ice cover, the effects of the winds are reduced because the stress is not transmitted to the water. The effects of seasonal changes of ice cover are also significant for wind stress surges in the Beaufort Sea in the Canadian Arctic.

Argentina and Brazil

Sea levels in the Rio de la Plata, between Argentina and Uruguay, are often influenced by strong southeasterly winds called 'sudestada'. In these cases the increased estuarine levels act as a barrier for the discharge from the Paraná and Uruguay Rivers. Flood levels are especially high when the sudestada coincide with high flows from these rivers; because of the extensive shallow and narrowing estuary, the surge standard deviations at Buenos Aires (Table 6.2) are unusually large. Further north the propagation of cold fronts from the south of Brazil northwards along the coast generally causes a succession of local storm surges. The main atmospheric forcing is usually the long-shore component of the wind, with a 10-hour lag for maximum surge levels at the coast.

General

In addition to the areas described above many other regions are vulnerable to surges generated by tropical storms. These regions include the coasts of southern China and Hong Kong, the Philippines, Indonesia, Northern Australia and the Queensland coast. In all cases the surge levels and storm damage are very sensitive to the direction and speed of the storm's progress, so that no two storms have exactly the same effects. In many cases flooding risks have been increased in recent years by urban developments, roads and concrete surfaces, which increase the immediate flow from rivers, sometimes to coincide with surges from the sea generated by extreme weather events.

6.6 Seiches

Tide gauge charts, particularly those from ocean islands and places linked to oceans by a narrow continental shelf, often show high-frequency oscillations superimposed on the normal tidal changes of sea level. These oscillations, called *seiches*, are due to local resonant oscillations of the water in the harbours and coastal areas in which the tide gauges are located. The nature of the resonance is the same as described in Section 4.2.2. Oscillations typically have periods from a few minutes to half an hour or longer, and amplitudes of a few centimetres. Sometimes the oscillations can be much bigger than this if triggered by an extreme forcing event such as a tsunami (see Section 6.7). Seiches are more difficult to detect with modern gauges that sample and record levels digitally, at intervals of several minutes, rather than producing a continuous trace.

The periods of these resonant seiches depend on the horizontal dimensions and on the depth of the water in the bay or harbour. They range from less than a second for a teacup to many hours for the seas and oceans. We have already seen how quarter-wave resonance can give large tidal amplitudes in some parts of the world such as the Bristol Channel in England and Long Island Sound in the United States. Similar resonant amplification, at shorter periods, occurs for seiches, especially where the quarter-wave resonance shown in case (2) in Figure 4.4 can be excited.

An example of these local oscillations imposed on the main tidal signal is given in Figure 6.10 for the south coast of Tasmania, Australia. A detailed analysis of sea level records in a harbour often shows that two or more periods of oscillation are present, as different types of oscillation are excited. Similar oscillations have been observed at other islands in the Indian Ocean, including the Seychelles. Several Mediterranean ports, including Malta, have seiches. In the Falkland Islands the seiches are particularly strong because of the strong prevailing winds. Elsewhere,

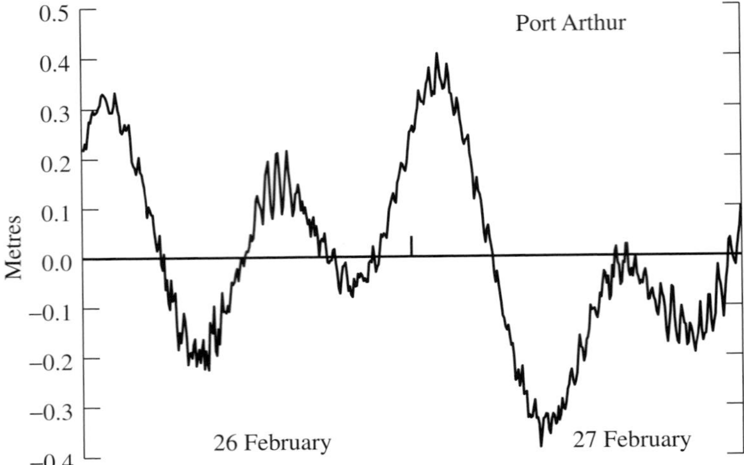

Figure 6.10. Twenty-minute seiches are clearly seen on top of this mixed tide at Port Arthur, Tasmania, Australia. The seiches occur in bursts, but the reason for this is not known (data supplied by John Hunter).

San Francisco Bay has oscillations with a period of about 45 minutes. Once established, these seiches may persist for several hours, showing that they are often only slightly damped by friction.

The reasons why seiching starts in any particular harbour or coast can be difficult to identify. There are several possible mechanisms: internal and surface gravity waves, winds, atmospheric pressure disturbances and seismic activity. To cause seiching the forcing must cause a change in the volume of water in the basin, through a flow of water in or out. Wind stress acting on the water in a harbour can cause gradients (Section 6.4.2) and if the wind relaxes there is a natural return of the water, which can set up seiches.

Some clues to forcing mechanisms can be found by looking at the way the amplitudes of the seiches change over time. In the Philippines, seiches have amplitudes that increase and decrease over a 14-day period, indicating that they are related to the astronomical spring–neap cycle. However, maximum seiche amplitudes lag the local lunar surface tide by about two days. This is because internal tidal waves generated some hundreds of kilometres to the southeast in the Sulu archipelago travel as internal waves until they reach Palawan Island. The internal waves (Section 4.5) travel at speeds of only about 8 km h^{-1}, taking about two days to reach the coast where they trigger local seiches. There are also seasonal variations in the lag due to changes in the internal stratification of the seas and hence in the speed of the internal wave travel. Because internal waves can have large amplitudes and a wide range of oscillating periods, they are very effective for generating seiches. Their importance in this context has only been confirmed in recent years.

Strong currents associated with seiches may reverse the tidal flows, and can disturb shipping by generating strong reversible flows in and out of a narrow entrance to a harbour, or by causing vessels to swing on their moorings. These rapidly reversing currents can also be important for coastal sediment movements and erosion: in a sense they are teeth on a saw which the tide moves up and down over the surface of the coast.

Where the energy in the seiches comes from external wave sources, the size of the entrance to an oscillating basin is critical; if it is large then much of the energy will be radiated away, but if it is too small then the external forcing, particularly if due to long period surface or internal waves, will not be able to feed energy into the oscillating system. A systematic analysis of the periods of seiches from a wide range of sites has not been made. It would be interesting to see whether there are preferred periods of seiching, which can be linked to natural internal resonances within stratified oceans, or preferred shapes of bays and estuaries.

6.7 Tsunamis

Tsunamis are rare wave events generally generated by seismic or other geological activity and as such fall outside the two principal categories of forces responsible for sea level changes: tides and weather. The term *tsunami* derives from the Japanese 'harbour wave', which is the form that they often take around the Japanese coast. An alternative name is *seismic sea waves*. The popular description of them as 'tidal waves' is incorrect because they lack the regularity associated with tides. Causes include sub-marine earthquakes, landslides into the sea and sub-marine slumping, for example of sediments on the continental slope. There are three distinct aspects of tsunamis that may be considered: their generation by earthquakes; their propagation in deep water; and their behaviour where they impinge on coasts and the surrounding regions of shallow water.

Not all sub-marine earthquakes produce tsunamis. The important element appears to be a vertical crustal movement which displaces the sea bed. After the sea bed is displaced, a tsunami is generated by the horizontal pressure gradients in the water, with gravity acting as a restoring force. The wave characteristics will depend on the amplitude of the displacement and the size of the sea bed which moves. Horizontal displacements of the sea bed will be relatively ineffective for producing tsunamis because water is not displaced vertically. Direct observations of sea bed displacements and their relation to tsunamis are not yet possible because these events are rare and inaccessible. It has been estimated that of the original energy released by a seismic event, only between one and ten per cent is transmitted to a tsunami.

Tsunamis that propagate across deep water typically have wave periods of ten minutes or longer, which means that their wavelength is long

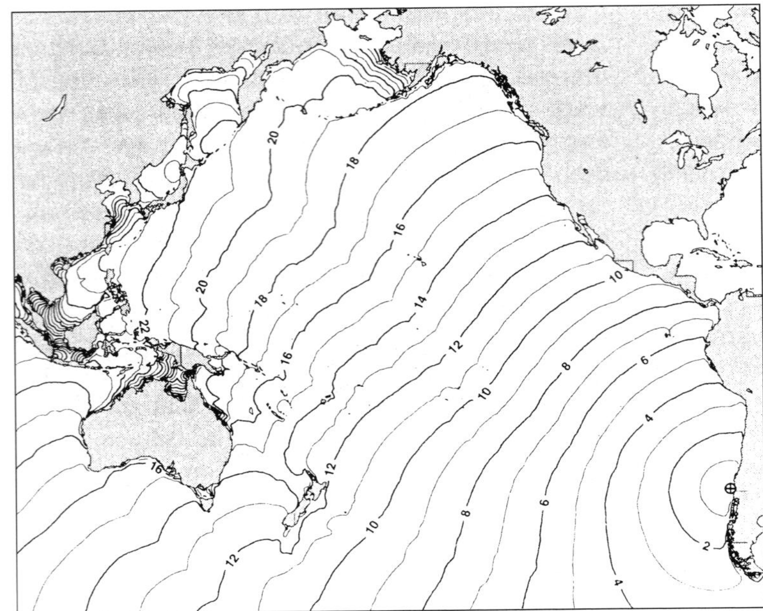

Figure 6.11. Travel times (in hours) for the 22 May 1960 Chile tsunami crossing the Pacific Basin. This tsunami was extremely destructive along the nearby coast of Chile, and caused significant destruction and casualties as far away as Hawaii and Japan (figure courtesy of Intergovernmental Oceanographic Commission of UNESCO).

compared with the water depth: their speed of propagation is therefore given by $(gD)^{\frac{1}{2}}$, as in Equation (4.1). For a wave of period of ten minutes and a water depth of 4000 m the speed is 715 km per hour, as for tides (Table 4.1), and the wavelength is 120 km. Amplitudes in deep water are small, probably not more than 1 m, so the waves pass unnoticed by ships at sea. Travel times for tsunamis between gauges in the Pacific Ocean confirm that the long-wave speed formula is valid, and this behaviour provides a basis for giving warnings to places further away from the source (see Figure 6.11 and the box Tsunami warnings).

An earthquake in the seismic zone along the coast of southern Chile generated the tsunami whose travel times are shown in Figure 6.11. This earthquake on 22 May 1960 was the largest ever instrumentally recorded. The Chilean government estimated 2 million people were left homeless and nearly 60 thousand houses were completely destroyed. At Chiloe Island near the earthquake epicentre the tsunami had a height of about 10 m, as it did at the towns of Valdivia and Corral where many people were drowned. A total of around 1000 people were drowned in Chile. Twenty-two hours after the original earthquake the wave reached Japan where local amplification caused tsunamis several metres high. Around 200 people were drowned in Japan although there had been plenty of time to warn them of the danger that was approaching.

Although the arrival time of a tsunami can be predicted accurately, the amplitude of the wave that hits a particular length of coast is much

Figure 6.12. Few photographs of tsunamis have captured the drama of a breaking tsunami as effectively as this historic 1946 image at Hilo, Hawaii. Scale is provided by the human figure illuminated. The devastation led to the setting up of the first tsunami warning system (photograph courtesy of NOAA).

less certain. This is because in shallow coastal waters, in addition to the normal amplification of the wave as it slows down, the wave undergoes reflection and refraction. Often the first wave is not the biggest. Sometimes the first arrival of the disturbance is seen as a recession of the sea, which exposes parts of the coast that have not been seen before. People who follow the water out to inspect the sea bed are then overwhelmed by the arrival of the main wave as a 'wall' of water 2 m or more high. Figure 6.12 shows an historic and dramatic example.

Depending on the nature of the original earthquake, the first arrival may also manifest itself as a sudden rise of sea level. Figure 6.13 shows positive first arrivals of a tsunami at three gauges in the northeast Pacific Ocean, following the Kuril Island tsunami of 4 October 1994. The record from the bottom pressure gauge in the deep ocean gives the earliest indication of the approaching tsunami.

The oscillations in Figure 6.13, which are superimposed on the normal tide, have different amplitudes and periods at each station, showing that they are local seiche motions (see Section 6.6). Each has characteristic local seiche periods initiated by the tsunami disturbance, rather than the periods being imposed by the tsunami itself. Once established the more extreme of these motions may continue for several days.

The Pacific Ocean and its coastlines are most vulnerable to tsunamis because of the seismically active surrounding plate boundaries. Hawaii experiences a significant tsunami on average once every seven years. An earthquake some 200 km off the coast of northeast Japan resulted in over 17 000 people being drowned in 1896. More recently, an earthquake on the evening of 7 July 1998 of magnitude 7.1 on the Richter scale

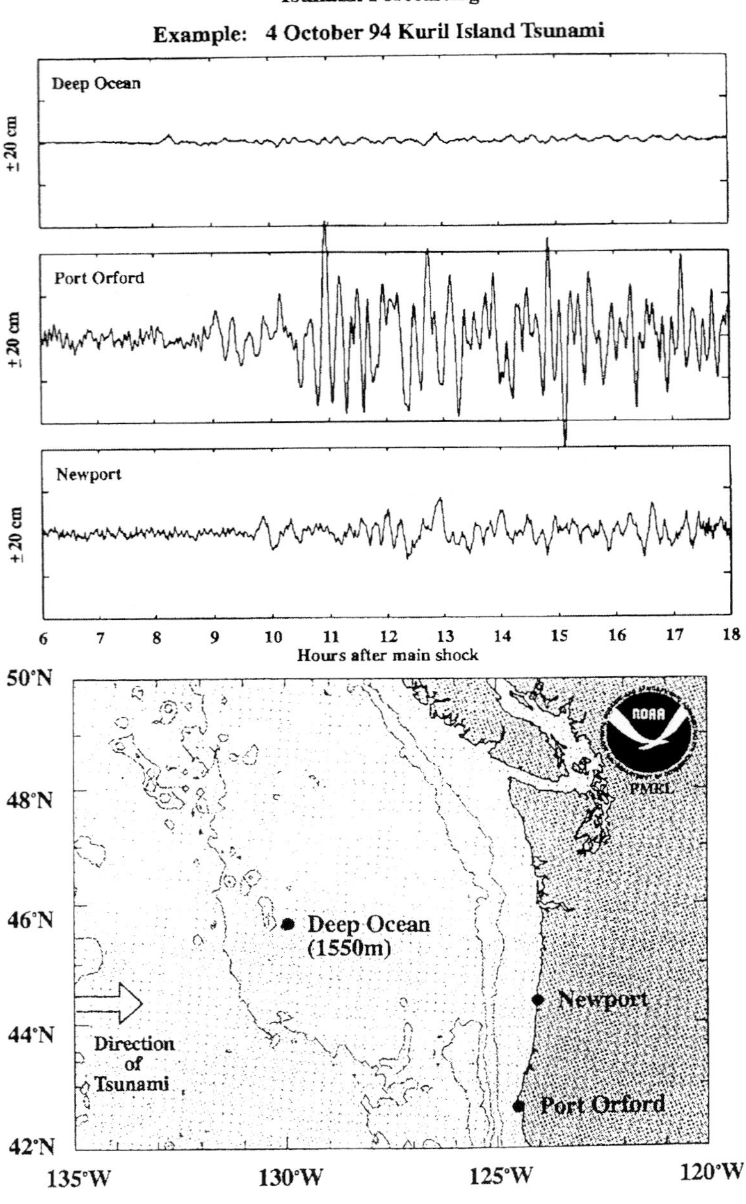

Figure 6.13. Three records of the arrival of the 4 October 1994 Kuril Island tsunami. The tsunami is detected (but very small) in the deep ocean. It shows that there can be large differences between signals of the same tsunami recorded on closely spaced shore-based gauges (figure supplied by NOAA and IOC).

near the northwest coast of Papua New Guinea was followed by a series of three catastrophic tsunami waves that devastated several villages along the coast killing more than 2000 people.

Tsunamis are much less common in the Indian and Atlantic Oceans, but when they do occur, they can be devastating. The Krakatoa explosion of 27 August 1883 was perhaps the most severe in recent history; more than 36 000 people were reported drowned by the tsunami which it generated in the Sunda Strait between Java and Sumatra. Its effects were observed not only all around the Indian Ocean, but also as far as South Georgia and Tierra del Fuego in the South Atlantic Ocean.

In the North Atlantic Ocean the most devastating recorded tsunami followed the Lisbon earthquake in Portugal on 1 November 1755. The direct destruction by the earthquake was followed by three enormous waves that overflowed the banks of the River Tagus. Seiches occurred in British lakes and harbours within a few minutes of the earthquake due to the direct seismic wave travelling in the solid earth; the main tsunami, travelling as a long progressive wave in the ocean, arrived on the coasts of the southwest of England and Ireland later in the afternoon. At Newlyn the sea level rose by more than 3 m in ten minutes.

There is no way at present of predicting the onset of a tsunami but warning systems have been established for the Pacific (see the box in Tsunami warnings) and are being considered for other regions at risk such as the Caribbean.

Flood warning systems

Defences and warning systems have been set up for cities and other low-lying areas at risk of flooding. Figure 6.14a shows the barrier that can be closed to protect part of the Netherlands. A similar barrier system in the lower reaches of the River Thames (Figure 6.14b) protects London from high sea levels in the North Sea. External surges travelling south along the east coast of Britain are assimilated into computer models in conjunction with forecasts of winds and air pressures to give warnings. Early 'Alert' warnings are issued and if a flood event becomes more probable these are followed by a 'Danger' warning, at which stage defensive actions are taken. Advance warnings of several hours are possible because of the way in which surges travel as Kelvin waves along the coast of eastern Britain from north to south, and are then reflected northwards along the coast of mainland Europe in the same sense and at the same speed as a tidal Kelvin wave (see Figure 4.10).

Along the east coast of the United States, NOAA operates SLOSH (Sea, Lake and Overland Surges from Hurricanes), a computer model, to give high-resolution levels of flooding. SLOSH operates over a range of 33 basins along the east coast from Laguna Madre, near the Mexican border in the south, to the Gulf of Maine in the north. The National Hurricane Center uses SLOSH, which is being continually improved, to estimate storm surge heights resulting from historical, hypothetical or predicted hurricanes, taking into account the hurricane pressure, size, forward speed, track and winds. Calculations are applied to each basin taking into account local features. Forecasts for actual events are updated every 6 hours.

Exact forecasting of flooding due to hurricanes is not possible because the exact track of the hurricane is critical; if the track forecast is inaccurate, then the forecast flooding levels will be wrong. Extra-tropical storm surges are forecast by an extension of the SLOSH model that incorporates dynamic forecasts of winds and air pressures. Using observations of levels and weather systems by continuously assimilating them in the models makes the forecasts much more accurate and useful.

Figure 6.14. (a) The Oosterscheldedam, the final section of the Dutch Delta Works system of sluices in the Netherlands (photo Copyright, The Ministry of Transport, Public Works and Water Management, Netherlands). (b) The Thames Barrier protects London from extreme sea levels (copyright the Environment Agency).

Tsunami warnings

After the 1960 Chilean earthquake and the ensuing tsunami disasters, an international warning system was set up for the Pacific Ocean, based in Hawaii. When an earthquake is detected on a seismograph, the tide gauges in the region are monitored to see whether a tsunami has been generated. As explained, only earthquakes that produce vertical movements of the sea bed cause tsunamis. If a tsunami is detected then estimated times of arrival can be given for each location in the Pacific. Tsunami travel time charts have been prepared for different earthquake and impact locations. Great emphasis is placed on educating the public so that they take tsunami warnings seriously and act accordingly. The United States west coast and Alaska are served by a local warning service which was set up after the large Alaskan earthquake in March 1964. This organisation responds by issuing tsunami warnings on average within ten minutes of an earthquake. Typically it is necessary to issue a warning once every two years. Tsunami warning systems maintain active websites and issue educational material. A similar warning system has been proposed for the Caribbean Sea and the Gulf of Mexico.

Further reading

More detailed accounts of the effects of air pressures and winds on sea levels may be found in Proudman (1953), Gill (1982), Jones and Toba (2001) and in many other physical oceanography texts. The data for Figure 6.5 is from Ross (1854); earlier tests of the inverted barometer effect are included in Lubbock (1836). Complications in the direct inverted barometer response are discussed by Mathers and Woodworth (2001). Heaps (1967) gives an early systematic account of the development of computer modelling and the forecasting of extra-tropical surges; Murty (1984) is a compendium of early surge publications. Flather (2001) gives a concise summary of the science of surges and an extensive account of flood forecasting in Flather (2000); see also Aikman and Rao (1999). Several Internet sites give information on systems for giving flood warnings, notably those maintained by NOAA. The science of forecasting tropical cyclones is covered in WMO (2002). Giese and Chapman (1993) and Chapman and Giese (2001) are both valuable reviews of seiches; the latter includes references to the seiching at Palwan Island, Philippines. Bryant (2001) gives a recent general account of tsunamis.

Questions

6.1 For a normal distribution of non-tidal residuals, 67 per cent of all values fall within one standard deviation of the mean. 95 per cent fall within two standard deviations and 99 per cent within three. From Table 6.2, assuming a normal distribution, what levels of non-tidal residual would be exceeded for 5 per cent of the time at Buenos Aires and at Newlyn? What percentage of non-tidal residuals would exceed 12 cm at Mombasa?

6.2 A low-pressure system is travelling over a continental shelf of 100 m depth, at 114 km h^{-1}. What is significant about this speed?

6.3 What would be the effect on local sea levels of the tropical daily air pressure changes described in Section 6.3, for a location on the equator and on the Greenwich meridian? Assume an instantaneous response to local air pressures.

6.4 How much higher would sea levels be at the head of a gulf 100 km long and 10 m deep, if they fully adjusted to a Strong Gale (22 m s^{-1}) blowing up the axis of the gulf?

6.5 An external surge is detected on a tide gauge at Aberdeen. Can you use Figure 4.10 to estimate how long it would take to reach Immingham?

Chapter 7
Mean sea level

In this chapter we deal just with mean sea level (MSL) changes as observed over the past 200 years or so, and the reasons for these changes. The strong public interest in global climate change often focuses on the possibility of rising global MSLs and the increased risks of coastal flooding. As we know from earlier chapters, increased risks of flooding will depend on changes in the tide and the surge characteristics of a coastal area, not just on increases in MSL. In Chapter 8 we will look at the problems of flooding risks and how these risks might change in the future, taking into account all three factors. Many of the processes involved have been introduced in earlier chapters, but here they are applied to changes over longer periods.

Historical tide gauge records are a rich source of information, but the location of these gauges is not optimum for MSL studies and, as shown in Figure 7.1, very few records go back more than 100 years. Most long-term gauges were originally placed around the coasts of highly developed countries for port operations, so most are in the northern hemisphere, particularly around the North Atlantic. Also, measurements are concentrated in estuaries and areas of intense human activity. Studies of changes in MSL now benefit from systematic wide-scale global measurements by satellite altimetry (Section 1.4.4); these records, although at present short compared with traditional measurements from tide gauges at fixed locations, will be increasingly important in the future for identifying long-term global and regional MSL trends.

MSL is defined as the average level of the sea surface, measured relative to a fixed level on the land, identified by a network of fixed benchmarks (Section 1.4.1). It is usually calculated as the average over a

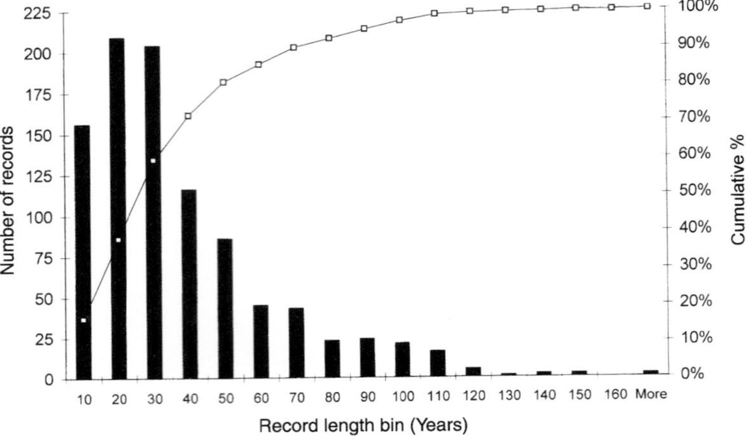

Figure 7.1. This histogram shows that most sea level records are less than 30 years long, which makes the detection of long-term trends difficult. There are very few records longer than 100 years (from Douglas *et al.* (2001), with permission).

specified long period such as a month, a year, or even over an 18.6-year nodal cycle. This averaging removes waves, all but the very long-period tides and storm effects.

It is often convenient to work with annual values of MSL. Averaging over a year is in a sense arbitrary, but it has the advantage that seasonal changes are removed; all systematic effects, including air pressure and winds, which persist for longer than a year are included.

Most countries have maps that refer their land levels to MSL, a practice that began during the nineteenth century. As we show in this chapter, this rough approximation to a global horizontal levelling datum or geoid (Section 2.5) is generally valid to better than a few tens of centimetres. Studies of changing sea level now work to a much higher accuracy, and the original levels used to define datums have changed. For example, the MSLs at Newlyn (1915–21) is defined as the Ordnance Datum for the levelling of Great Britain. However, present-day MSLs at Newlyn are more than 0.10 m above the earlier datum level.

Figure 7.2 shows the annual MSL variability for some of the longest tide gauge records held by the international Permanent Service for Mean Sea Level (PSMSL). There are long-term trends and inter-annual changes in all the records.

The preferred units for describing MSL changes are mm yr^{-1}; some authors use cm century^{-1}. Note that for palaeo sea level change rates, geologists often use metres per thousand years, which is numerically the same as mm yr^{-1}. For studying long-term MSL changes it is essential that the tide gauge benchmarks to which they relate are stable and regularly checked.

We start this chapter by looking at how MSL is calculated from sea level records. We then consider how MSL differs spatially from the

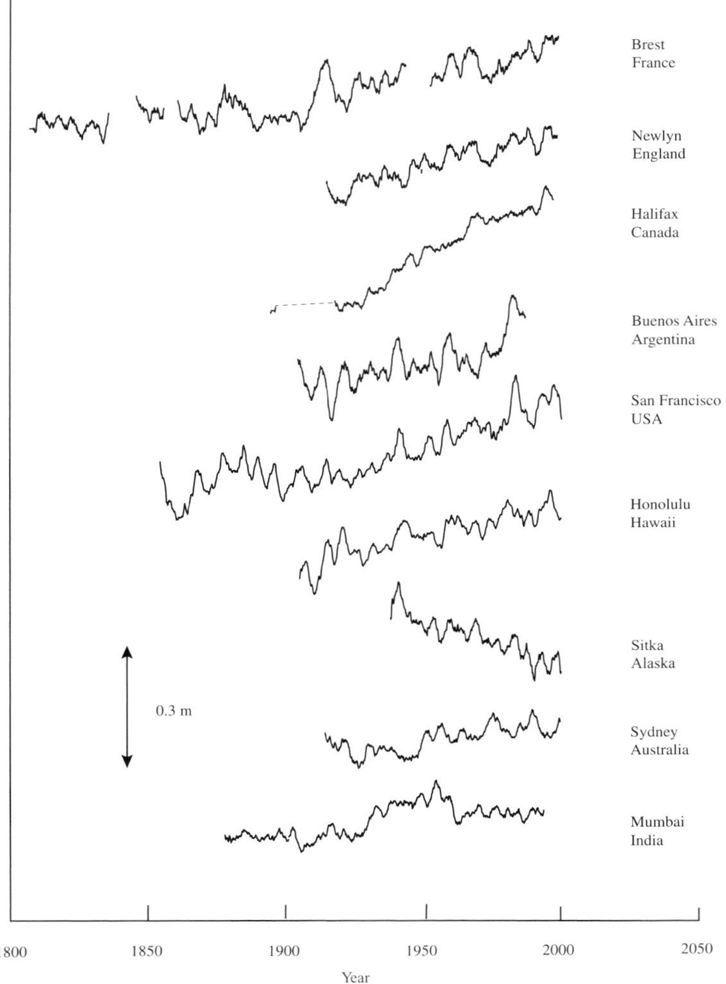

Figure 7.2. Some of the longest records of annual mean sea levels from the Permanent Service for Mean Sea Level. The values plotted are three-year running means. There is a general, but by no means universal, upward trend. Brest and Newlyn are highly correlated, and there is some correlation among the records from the Pacific Ocean.

geoid around the globe. This leads to a discussion of the variations of MSL at fixed locations over years and decades. We look in more detail at long-term trends of MSL changes and their causes: ocean thermal expansion and the melting of ice, including the important residual effects of earlier ice age glaciation episodes. This prepares us for Chapter 8, which deals with all aspects of coastal flooding and climate change, including estimates of future MSL rise.

7.1 Calculating mean sea level

There are several ways of calculating MSL in order to eliminate the unwanted short-term changes of relatively large amplitude due to tides and surges. The most direct way is to sum all the values observed over one month and calculate the arithmetic mean. The annual mean can be calculated in the same way. More elaborate techniques eliminate short-term changes by applying a low-pass digital filter before taking the arithmetic average. This avoids the influence on the MSL of including part of a tide in the averaging period. It is now normal practice for national authorities to calculate MSL for each month and year, using sea levels observed every hour. If more frequent observations are available (typically six- or fifteen-minute values), then these can be used in the same way, directly or after numerical filtering to give hourly values.

Before computer processing reduced the labour of calculations, some authorities found it easier to determine the mean tide level (the average of all the high and low waters) within a specified period. Indeed, in many places in the early days of sea level measurements, only high water and low water levels were recorded. Mean tide level (MTL) is not the same as MSL because of the influence of the shallow-water tidal harmonics, particularly the lunar fourth-diurnal term, $\mathbf{M_4}$. Figure 5.3 shows an extreme example of uneven high and low water levels due to shallow-water distortions. Typically the difference is not more than a few centimetres. Some of the old MTL values should be adjusted for this difference, for comparison with MSL. This is done by looking in detail at the phases of the higher tidal harmonics (see Chapter 5), especially $\mathbf{M_4}$, in relation to the main tidal constituents; for a tide of just $\mathbf{M_2}$ and $\mathbf{M_4}$, in the worst case where the two constituents are both at their maximum at high water, MTL would exceed MSL by the amplitude of $\mathbf{M_4}$.

We can now also calculate global MSLs using data from satellite altimetry. The arithmetic average of sea level measurements from a satellite over a complete cycle of orbits gives a global average over a specified period. This then represents signals due to total ocean volume fluctuations, whether they are due to changes in ocean mass or thermal expansion of a fixed mass. Figure 7.3 shows ten-day and sixty-day smoothed values of the global MSL from TOPEX/Poseidon after removing the annual and semi-annual variations and correcting for instrument effects using a global network of tide gauges for calibration checks. The importance of removing satellite bias and drift by comparison with a high-quality tide gauge network was discussed in Section 1.4.4. The combination of long-term datum stability from the tide gauge network and GPS fixing, together with the global coverage and precision of altimeters, provides an integrated, long-term and robust system for measuring global MSL changes.

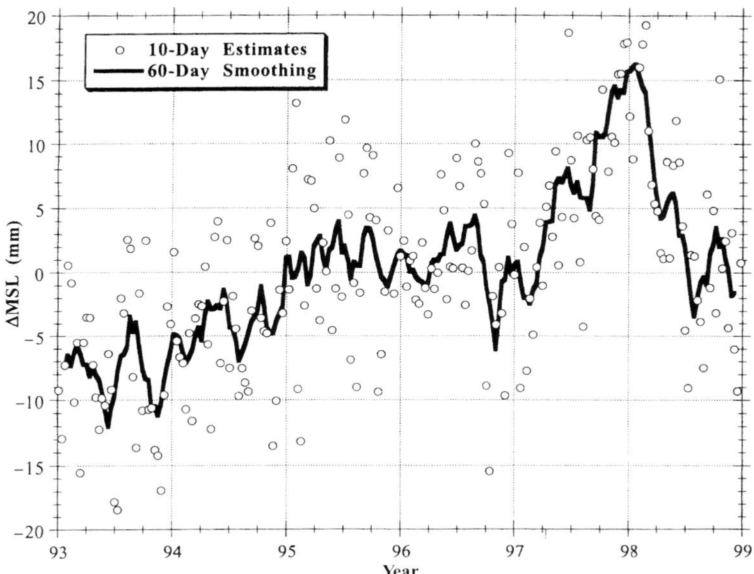

Figure 7.3. Global sea levels averaged over a complete ten-day cycle of TOPEX/Poseidon (open circles). The solid black curve shows averaging over sixty days. The high values in 1997–98 are attributed to an exceptional El Niño event. Annual and semi-annual variations have been removed (from Fu and Cazenave (2001), with permission).

7.2 Spatial changes in mean sea level

On the geoid surface there are no horizontal forces to disturb the water because the force of gravity is always perpendicular to this surface. The sea surface would adjust to take the form of this gravitational equipotential surface (Section 2.5) if there were no other effects. This means no tidal forcing (no moon or sun), no spatial differences in water density (no solar heating or precipitation differences to generate water density differences), no atmospheric pressure differences, no winds and no currents. In reality these effects are always present. There are permanent MSL surface deviations of as much as a metre or more from the geoid. These differences are related to ocean circulation and dynamics. They can be measured by satellite altimetry (Section 1.4.4), after subtracting the geoid surface.

Our knowledge of the geoid surface is based on gravity *in situ* measurements and satellite orbit analysis; in the future it will be much improved by space gravity missions dedicated to geoid mapping (Section 2.5). These excursions of MSL from the geoid are roughly two orders of magnitude less than the geoid deviations from a regular ellipsoid of revolution. Some scientists call this the mean sea surface (MSS), reserving the term MSL for time changes at a particular location. We include both spatial and time changes in our definition of MSL.

Figure 7.4 shows how, in the real oceans, the sea surface is displaced from the geoid, usually by a few decimetres but sometimes by more

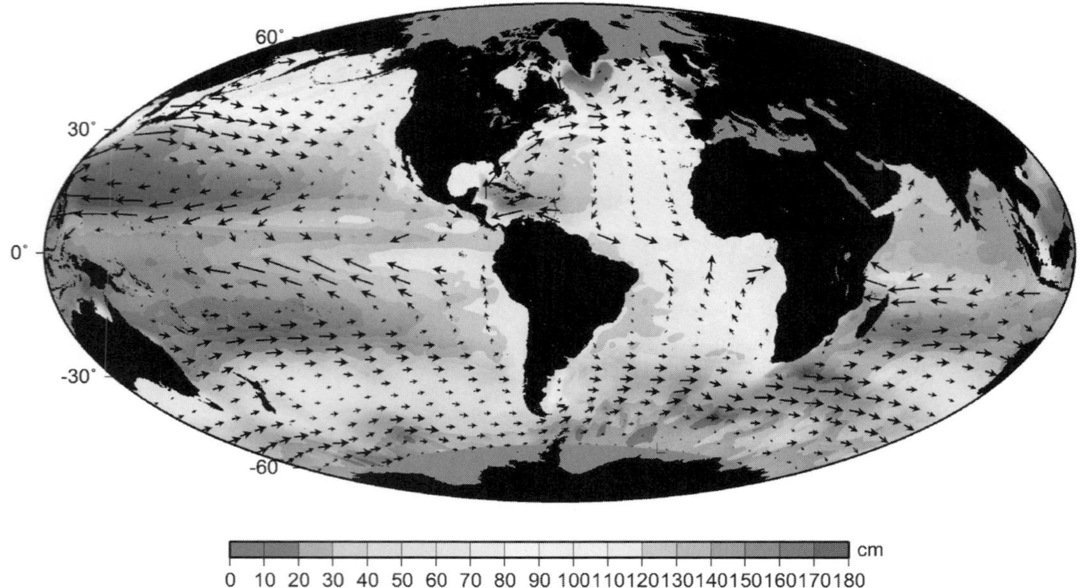

Figure 7.4. This map shows how the MSL varies from the geoid because of ocean density, currents and meteorological effects. The greatest differences are found across ocean currents (black arrows). Zero is the lowest level (courtesy Aviso-CLS). See colour plate section.

than a metre. The greatest surface slopes and differences are found in the areas of the great ocean currents (Gulf Stream, Kuroshio, Antarctic Circumpolar Current). There is a close dynamic relationship between the currents and the slope of the MSL surface relative to the geoid. As in the case of a Kelvin wave (Section 4.2.3), there is a slope of the sea level up to the right of the current in the northern hemisphere (to the left in the southern hemisphere). This slope provides the force to counteract the earth's rotation. However, in the present case, for a steady current the relationship is much simpler:

$$vf = g \times \text{slope} \tag{7.1}$$

where v is the current speed, f is the Coriolis parameter ($f = 2\omega_s \sin\phi = 1.459 \times 10^{-4} \sin\phi$, where ϕ is the latitude, see Section 4.2.3) and g is gravitational acceleration. Slope is as defined in Equation (6.3); here it is measured at right angles to the direction of the current v and is positive upward to the right of the direction of the current in the northern hemisphere. This simple relationship is called the geostrophic balance and it is fundamental for computing ocean circulation in physical oceanography.

Distributions of MSL (or MSS) such as that shown in Figure 7.4 are calculated using the traditional methods of physical oceanography, knowing the temperature and salinity distribution in the ocean. These distributions of MSL are used to compute currents. The calculations are based on the assumption of a deep layer in the ocean where there are no currents and no horizontal pressure differences. The constant pressure on this surface is expressed by Equation (1.2): it means that, ignoring air pressure differences, an increase in the average density (lower temperatures, higher salinities) of the water column results in a lower sea level, and vice versa. The level of 'no current' is not strictly valid anywhere in the ocean and so the surface shown in Figure 7.4 is only an approximation to the true MSL surface. The best way to map the true MSL surface is to use altimetry, with adjustment for the geoid (which still needs to be more accurately determined), as discussed above.

In Figure 7.4 the lowest MSL values are found off Antarctica. The highest MSL values are found in the tropics and particularly in the tropical Indian and Pacific Oceans, where the warmer water leads to lower average densities in the vertical water column. Because of these *steric level differences* in the two oceans, Pacific Ocean MSL is some 0.20 m higher than Atlantic Ocean MSL, at each end of the Panama Canal.

MSL is also influenced by permanent wind and pressure fields (Sections 6.3 and 6.4). Part of the MSL variation from the geoid shown in Figure 7.4 is due to variations in the mean atmospheric pressure and associated average winds. The convergence and divergence of ocean surface currents due to wind-driven Ekman transport (Section 6.4.3) leads to western boundary currents such as the Gulf Stream in the North Atlantic and the Kuroshio current off Japan. As explained, to maintain the geostrophic balance, these currents have permanent MSL gradients across their axis of flow (Equation (7.1)), because of the rotation of the earth and the need to provide a force to compensate for the rotation.

More local variations in average water density, currents, air pressures and winds will also be seen as permanent local spatial variations in the local MSL from the geoid.

7.3 Observed annual and inter-annual changes

In addition to the permanent spatial variations of MSL from the geoid discussed in the previous section, there are changes at all locations over months and years. The main causes of MSL variability over a year or longer periods (inter-annual changes) are the direct and indirect effects of varying climate, particularly the inter-annual changes in winds and solar heating on ocean currents.

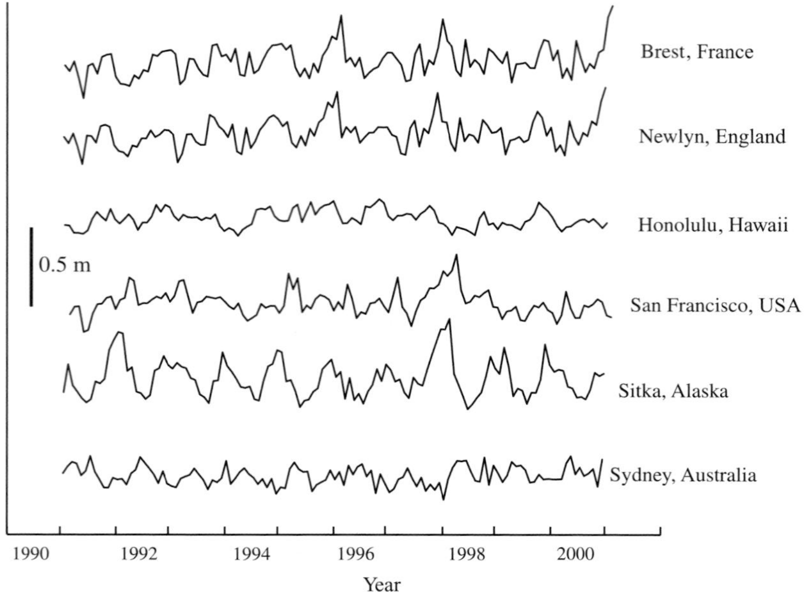

Figure 7.5. Ten years of MSL data from selected sites showing monthly variations in more detail. Again, Brest and Newlyn are highly correlated. The strongest seasonal cycle is seen at Sitka, Alaska (data from PSMSL).

7.3.1 Seasonal (annual) changes

Careful examination of Figure 7.5, which shows monthly MSL over a ten-year period at six long-established tide gauges, reveals a seasonal variation in MSL. Typically, this annual or seasonal cycle of sea level has an amplitude of 40 to 70 mm in mid-latitudes, but it is often masked by other variability; in the tropics the annual cycle is much reduced. Coastal gauges may show larger seasonal ranges, particularly if they are affected by seasonal river discharge. Analyses of Pacific sea levels have shown that seasonal amplitudes in excess of 100 mm occur in the East China Sea and along the coast of southern Japan, whereas amplitudes of less than 25 mm are found in some ocean island groups such as the Marshall Islands and Samoa.

Analysis of altimetry data shows the seasonal signal very clearly and also shows how it is out of phase between the northern and southern hemispheres. In the northern hemisphere sea levels generally reach a maximum value around September, at the end of the summer heating and expansion. Along the Pacific coast of North America, the month of maximum seasonal MSL increases systematically from September in the south to December or even January in the north. Conversely, in the southern hemisphere the seasonal cycle, which is usually smaller than in

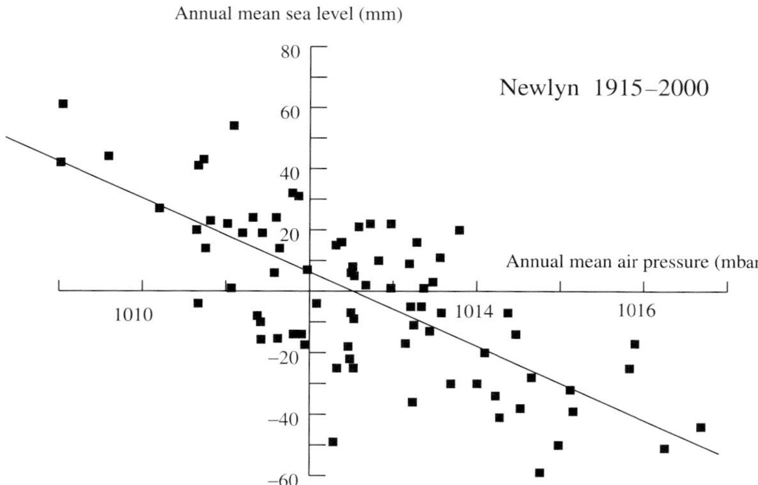

Figure 7.6. Annual MSLs at Newlyn are affected by annual mean air pressures according to the inverted barometer effect. Here the correlation is –11.9 mm mbar^{-1} of air pressure increase. The difference from the theoretical value is due to correlation between air pressures and winds. Trends and MSL have been removed.

the north, reaches a maximum around March; generally the maximum values in Australia and New Zealand occur slightly later, from April to June. The smaller amplitude of the seasonal cycle in the southern hemisphere has been attributed to the smaller area of continental mass.

7.3.2 Air pressure effects

In Section 6.3 we saw how sea level and air pressures are related through the inverted barometer effect. The same effect occurs for annual MSL values: if the average air pressure is high for a particular year, then the sea level will usually be lower. Figure 7.6 shows this relationship for Newlyn, where the response of MSL (-11.9 mm mbar^{-1}) is very close to, but not exactly, the theoretical -10.0 mm mbar^{-1}. Long-term trends in MSL can be seen more easily if the annual MSL values are adjusted for this air pressure effect.

7.3.3 Ocean circulation

Inter-annual changes of MSL are sometimes coherent over large distances, for example between San Francisco and Honolulu in the period 1920–50 (Figure 7.2). The longer term changes shown are due to slow changes in steric levels, winds and associated ocean circulation. The scientific study and description of ocean circulation is beyond the scope of this book (see the Further reading section) but we can look at a few examples of how changes in ocean dynamics affect MSL.

In the tropical Pacific Ocean El Niño, the air–sea coupled oscillation on inter-annual time scales has a clear MSL signature. The warm

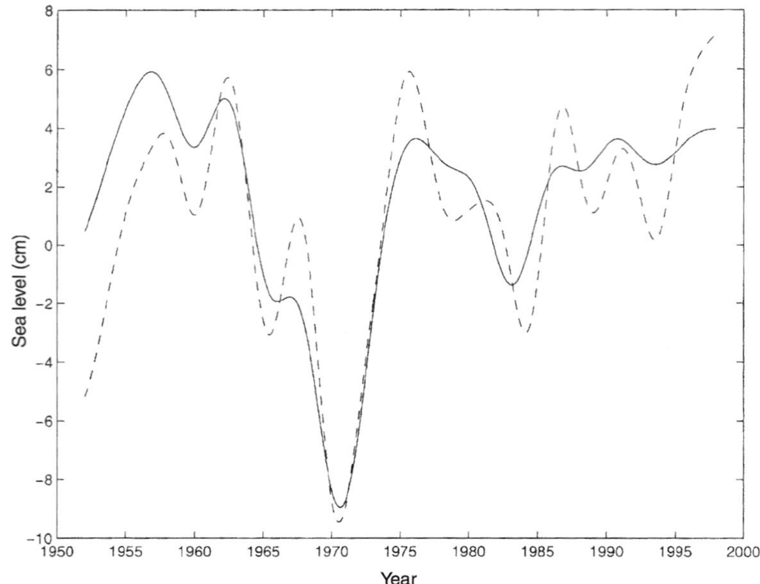

Figure 7.7. A comparison between the observed MSL at Bermuda (dashed curve) and the MSL for Bermuda computed from Atlantic Ocean responses to winds (solid curve). The model is forced only by open ocean winds which here cause long-term sea level differences of more than 15 cm (from Douglas *et al.* (2001), with permission).

phase is always marked by a 10–30 cm rise in sea level in the equatorial eastern Pacific. Increased MSL, measured by altimetry, is an indicator of increased upper heat content in the tropical ocean. The 1997–98 peak in global MSL (Figure 7.3) is associated with higher heat content in the Pacific Ocean, due to the exceptional El Niño event at that time.

Changes in MSL along the western side of the Atlantic Ocean have been associated with low-frequency Rossby waves forced by large-scale winds over the ocean. Carl-Gustaf Rossby (1898–1957) was a pioneer in the development of theories of fluid movements on the rotating earth. We have already discussed the Rossby radius in Section 4.2.3. Rossby waves are a form of wave that depends on the change of the earth rotation Coriolis parameter f with latitude: in the ocean they propagate slowly and only towards the west, with a speed of a few centimetres per second. A wave takes about five years to cross the Atlantic, which results in quite different MSL variations on opposite sides of the ocean. Figure 7.7 compares observed MSL at Bermuda with a Rossby wave model calculation where the model is forced only by winds over the whole width of the Atlantic Ocean. This confirms that regional winds and sea levels are closely related on these longer time scales.

Another reason for sea level changes at a site due to ocean circulation is the passage of an ocean eddy. Open-ocean eddies are observed almost anywhere at mid- and high latitudes, especially where they spin off from western boundary currents. The energy in these eddies often exceeds the mean flow of the current by an order of magnitude. Typical eddies can have dimensions from 50 to 500 km and persist over periods from 10 to 100 days. Eddies are seen in altimeter data as time-varying (this distinguishes them from permanent geoid variations) sea levels of the order of 10 or 20 cm amplitude or more, depending on the intensity and scale of the individual eddy. As an eddy moves slowly past an ocean tide gauge, it is seen as a slow fall and increase in the MSL.

As our measurements with altimeters and our modelling of global ocean circulation become more reliable, the inter-annual MSL variations obvious in Figure 7.2 may be adjusted, giving better estimates of long-term sea level trends. This adjustment is an extension of adjustments for annual mean air pressure (Figure 7.6) that are already routinely made. Past MSL variations from archived tide gauge records may also be used to infer historical changes in ocean circulation, and there is much research work needed on this topic.

7.3.4 Nodal MSL changes

There is also a small periodic component of the Equilibrium Tide that has an 18.6-year period. This is the nodal tide (**N**), so called because it is due to the nodal effects on lunar declination described in Section 2.4.3. The Equilibrium form of **N**, which depends on the latitude, has an amplitude of 18 mm at the equator. Maximum amplitudes of the nodal MSL tide are indicated for March 1969, November 1987, June 2006 and at 18.6-year intervals thereafter. In practice amplitudes for **N** are very small compared with the background ocean variability considered in the previous section, and the nodal MSL changes are only found through detailed scientific analysis. Around Europe the nodal MSL changes have amplitudes of around 4.4 mm, less than expected from the nodal Equilibrium Tide. Small periodic changes in MSL have also been attributed to the Chandler Wobble of the axis of rotation of the earth with a period close to 436 days, and also to the 11-year sunspot cycle.

7.4 Isostatic adjustment

As Figure 7.2 shows, there is considerable variability in the long-term MSL trends observed relative to local tide gauge benchmarks. Figure 7.8 shows the very wide scatter of relative sea level trends for stations with records over twenty years or longer. In high northern latitudes, notably Sitka, Alaska and northern Scandinavia, there are large downward trends in sea level. Juneau, Alaska, has a downward trend of more than 1.0 mm yr^{-1} although it is barely 150 km north of Sitka, where the downward rate is only 0.25 mm yr^{-1}. Regional vertical crustal movements

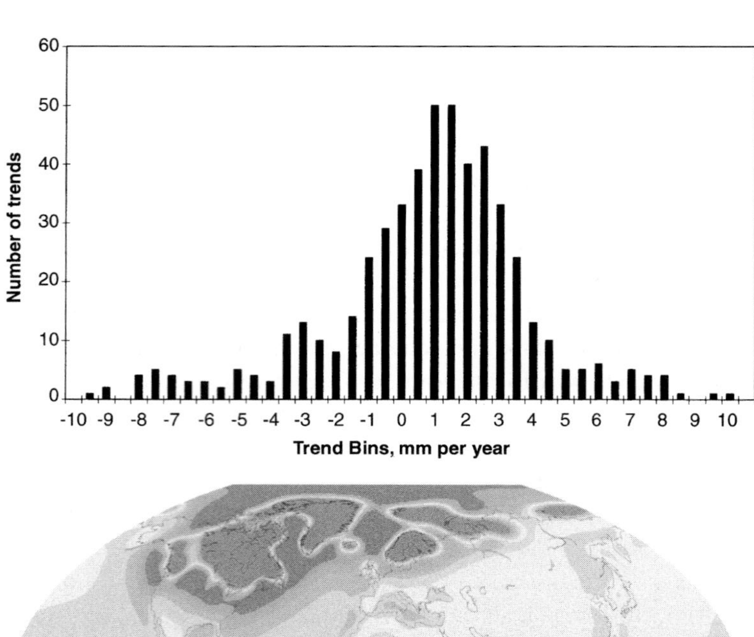

Figure 7.8. This figure shows that there is considerable variation in the local long-term MSL trends about the global average of 1–2 mm yr^{-1}. The differences are due to local geological effects, particularly post-glacial adjustment (adapted from Douglas *et al.* (2001), with permission).

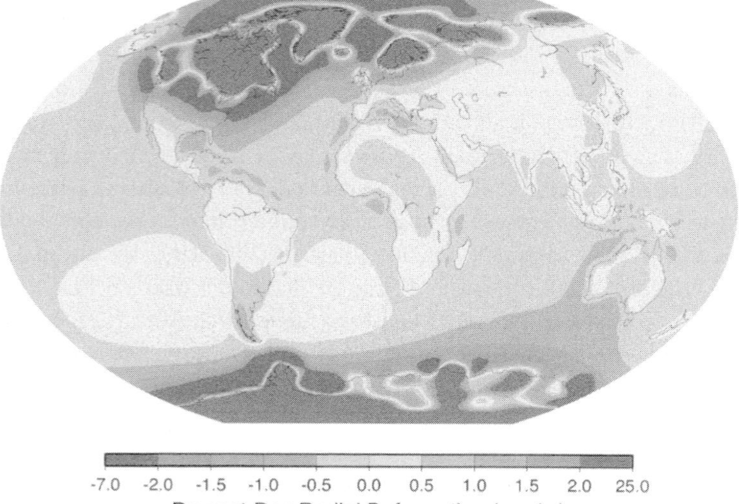

Figure 7.9. This map shows how the land is moving vertically through post-glacial adjustment after the removal of the ice sheets. It shows land is rising in the polar regions and generally falling in the tropics. The maximum uplift rates (red) reach 20–30 mm yr^{-1} in places (for example, northeast Canada). The maximum subsidence rates reach 6–7 mm yr^{-1} (for example, between Greenland and northeast Canada) (provided by Glenn Milne, University of Durham; see Mitrovica and Milne (2002)). See colour plate section.

in Alaska are complicated by frequent local earthquakes. Large downward trends of sea level are also observed in the northern Baltic Sea, particularly in the Gulf of Bothnia along the coast of both Sweden and Finland. These apparent downward trends of sea level are actually due to vertical upward land movements. Global patterns of these vertical land movements are shown in Figure 7.9.

One of the major difficulties in the study of longer term changes of MSL, called *secular changes*, is the separation of changes due to ocean effects from local land movements. Because MSL changes are measured relative to a fixed benchmark on land, measured sea levels include both, and the observed secular changes may be partly due to the movement of the fixed benchmark and partly to an ocean level change. Vertical land movements may be due to any of a wide range of geological factors: earthquakes; consolidation of coastal sediments such as deltas; consequences of extracting oil or water. The most general is the response of the earth to the melting of the high-latitude glaciers after the last ice age over a period from 5000 to15 000 years ago (Figure 7.9). This recovery from glacial loading is called Glacial Isostatic Adjustment (GIA) or sometimes Post-Glacial Rebound (PGR); the former term is now preferred because the adjustment is not everywhere upwards.

The response is greatest in high latitudes, as the earth moves up after the weight of ice has been removed. This explains the apparent sea level fall at Sitka. To compensate for the vertical uplift of land due to isostatic recovery from glacial loading the solid earth responds with internal mass movements: there is a surrounding region of the earth's surface which is correspondingly sinking. In this area, known as the *forebulge*, present-day sea levels appear to be rising more rapidly than elsewhere; these regions include parts of Nova Scotia, Canada, southeast England and the Netherlands.

Changes of sea level due to changes in the volume of water in the ocean, for example as a result of the melting of glaciers, are known as *eustatic* sea level changes. It is now accepted that eustatic rises cannot be observed free of residual post-glacial distortions, even in the tropics. Here it is because of downward isostatic vertical land movements over long periods, which compensate for the additional weight of higher sea levels pressing on the adjacent sea bed. Almost everywhere away from former ice sheets the GIA results in land sinking, so MSL trends from these sites must be systematically corrected upwards.

Some authors assume that eustatic changes will be seen as the same change of MSL everywhere, but this assumption is only a first approximation to the actual relative changes, which have considerable local variations, as shown in Figure 7.8 and Figure 7.9. Eustatic sea level rise cannot be observed free of residual GIA distortions, even on tropical coasts well away from the direct influence of polar ice caps.

Precise satellite navigation systems such as GPS can fix the position of a tide gauge benchmark in geocentric coordinates with an accuracy of around 10 mm. Over the next several decades the vertical movement of benchmarks fixed by GPS at tide gauges will be determined and, as this happens, local relative MSL measurements can be corrected for all local land movements from both natural and human causes (see Figure 1.11). It will be possible to detect routinely vertical crustal movements due to whatever cause, to better than a few millimetres per year, and to decouple them from the relative sea level changes measured by the tide gauge. We cannot do this very well at present because the GPS records are too short.

Many countries now have programmes of regularly fixing the benchmarks of important tide gauges. The best possible accuracy of GPS requires special corrections for atmospheric effects by using a dual-frequency (see Section 1.4.4) and the use of precise satellite orbit calculations. Corrections are also made for antenna phase centre variations and for local land movements due to earth tides and tidal loading at the time of measurement.

These GPS programmes will substantially reduce the range of sea level trends seen in Figure 7.8 and allow much more accurate estimates of the change in ocean volume and eustatic sea level variations. Reliable measurements of trends from GPS may take several decades, so meanwhile it is necessary to adjust relative MSL trends for glaciation using computer models of ice loading and the response of the earth's mantle, as shown in Figure 7.9.

Relative vertical land movement between two sites can sometimes be detected by plotting the difference between the two separate MSL time series (this is called 'buddy checking'). The effects of ocean dynamics tend to cancel, as they are normally coherent over a few hundred kilometres, and the residual differences are due to vertical land movements. Figure 7.10 shows the annual averages of sea level at Trieste

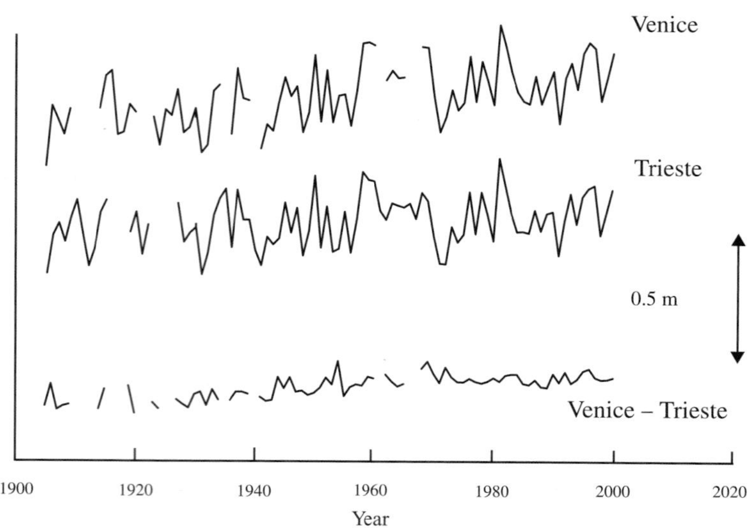

Figure 7.10. Trieste and Venice mean sea levels in the Adriatic Sea. These show similar short-term variations due to ocean effects. The bottom plot of differences shows a long-term divergence due to local vertical land movement. Venice was sinking relative to Trieste at the start of this period, but this divergence has now stopped (data from PSMSL).

and Venice, with coherent inter-annual variations of MSL. There is a long-term divergence in the differences after 1930, although the two cities are only about 100 km apart. The more rapidly rising sea level at Venice compared with that at Trieste shows a difference in the rate of vertical land movement; it has been suggested that the abstraction of groundwater from aquifers beneath Venice has led to compaction of sediments there (sinking land) and thus a more rapid rise of relative sea level.

Over periods longer than we can measure using tide gauge data, coastal geologists have looked for local indicators of long-term vertical land movements such as raised beaches, tidal flats and salt marshes as well as submerged wave-cut terraces, as indicators of GIA and other tectonic changes. The subject of vertical land movements and eustatic sea level rise is a fascinating and complex issue, which is discussed in detail in other publications.

7.5 Changes of water volume

Although, as discussed in the previous section, the eustatic rise of global sea level cannot be measured directly, it is still a useful concept. Eustatic changes of sea level can be defined as the change in seawater volume divided by the ocean surface area. The increase in ocean volume is due to two general factors: expansion due to warming, especially of water in the upper ocean; and an increase in mass of the ocean due to the melting of grounded ice. Floating ice, such as that covering the North Polar regions, will not affect sea levels if it melts because it already displaces its equivalent volume of water. This accords with Archimedes' Principle, which can be seen in a more direct way when the level in a glass of water (or gin) is unchanged as ice floating in it melts. Based on tide gauge data, the rate of global MSL rise during the twentieth century is in the range 1 to 2 mm yr^{-1}. Table 7.1 and Figure 7.11 summarise the estimated contributions to the overall sea level rise in the twentieth century based on IPCC (2001). We can consider the main factors in Figure 7.11 in turn.

7.5.1 Thermal expansion

Thermal expansion is one of the major contributors to observed sea level change. The process of heat absorption and ocean expansion is quite complicated: for a given heat input the warmer surface waters of the ocean expand more than the colder deeper waters, which are near to their temperature of maximum density. Recall that freshwater has a maximum

Table 7.1. *Estimated components of rates of sea level rise from observation and models averaged over the period 1910–90 (from IPCC (2001)).*

	Minimum (mm yr^{-1})	Central (mm yr^{-1})	Maximum (mm yr^{-1})
Thermal expansion	0.3	0.5	0.7
Glaciers and ice caps	0.2	0.3	0.4
Greenland–twentieth–century effects	0.0	0.1	0.1
Antarctica-twentieth–century effects	−0.2	−0.1	0.0
Long-term ice sheet adjustments	0.0	0.25	0.5
Permafrost	0.00	0.025	0.05
Sediment deposition	0.00	0.025	0.05
Terrestrial storage	−1.1	−0.35	0.4
TOTAL	0.8	0.7	2.2
Estimated from observations	1.0	1.5	2.0

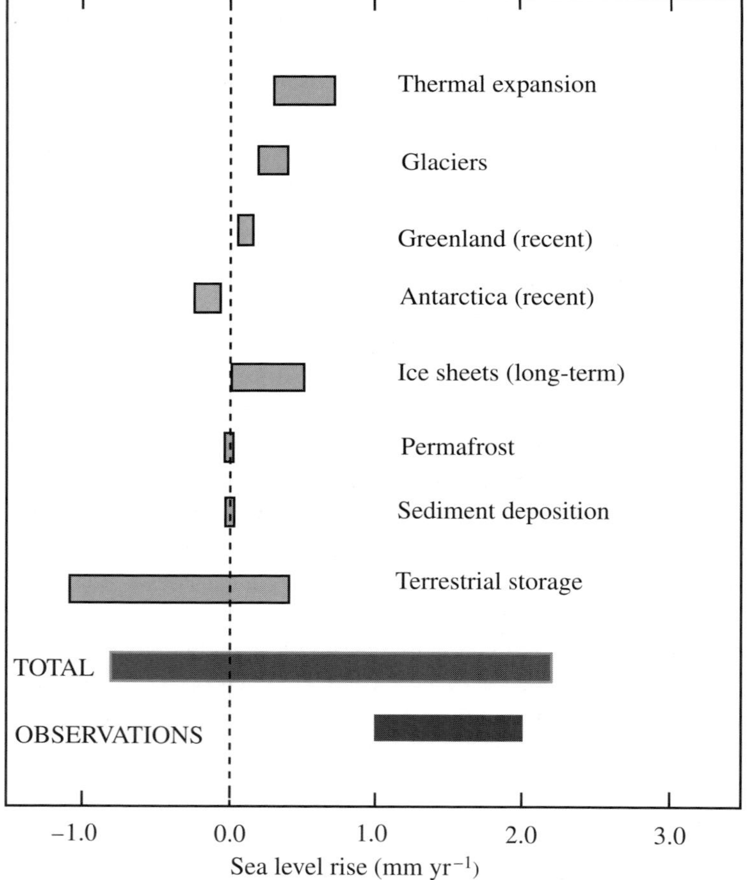

Figure 7.11. A summary of the various effects that contributed to MSL changes in the period 1910–1990. Widths show the uncertainty in each individual estimate (from IPCC (2001), with permission of Cambridge University Press).

density at 4°C, so adding or removing heat at 4°C will not change the volume or density. The dissolved salt changes the properties of seawater from those of freshwater, so seawater does not have a temperature of maximum density, but at the low temperatures of the deep oceans the thermal expansion is still small. Therefore water will expand more in the warm tropics than in the colder oceans at high latitudes, for the same temperature change. This means that the way in which the oceans take up heat affects the overall expansion, and also leads to regional differences and delays.

The delay in the MSL adjustment to changing global temperatures is an important aspect of the complex process. The large heat capacity of the oceans means that there will be a delay before they reach thermal equilibrium with the new temperatures and the full effects of global warming are evident. For this reason increases in atmospheric temperatures will continue to cause sea levels to rise for centuries after they occur. In seeking to explain present rates of MSL change, we must take into account changes in air temperatures over several past centuries.

Long-term changes are difficult to detect because there are very few long-term ocean temperature series available. The most convincing evidence for water warming is from the North Atlantic, one of the best-measured and studied ocean areas. On decadal time scales the variations in sea levels at Bermuda correspond well with the temperature changes at a local long-term deep monitoring site, 'Station S', over more than 70 years, and ocean sections show the temperature increases are widespread (see Section 7.3.3). Overall, there are indications of thermal expansion, particularly in the sub-tropical gyres, equivalent to sea level changes of the order of millimeters per year. Although the evidence of thermal expansion is most convincing for the North Atlantic, it also applies to the Pacific and Indian Oceans. The only areas of ocean cooling appear to be the sub-polar gyres, particularly in the North Atlantic Ocean, but measurements there are sparse.

An alternative approach to estimating MSL rise due to thermal expansion from direct observations is through ocean modelling. Several different kinds of computer ocean–atmosphere models show a general increase in ocean volume in response to the observed increases in air temperature over the past century. They suggest that the average rate of sea level rise due to thermal expansion was of the order of 0.3 to 0.7 mm yr^{-1}, increasing to 0.6 to 1.1 mm yr^{-1} in recent decades. There is a tendency for models to show a lower than average increase of sea level around the Southern Ocean. In general the detailed geographic sea level changes are not yet in good agreement among the models, nor with the sparse measurements of temperatures and sea level change.

7.5.2 Melting ice

Over geological time scales there have been substantial glacial episodes that have changed MSL by more than 100 m. The most recent glacial maximum reached a peak around 20 000 years ago, when MSL was probably 120 m lower than today. Present levels were reached some 6000 years ago; since then there have been significant though smaller changes, with many local variations. The present mass of water in the oceans is controlled mainly by a balance between processes of ice formation and melting.

The melting of glaciers and small ice caps has been well documented in many places. The total water contained in ice caps and glaciers (excluding the ice sheets of Antarctica and Greenland) is equivalent to about 0.4 m of sea level. Glaciers and ice caps are rather sensitive indicators of climate change. Their mass balance, while well known for a few glaciers, is not well documented for most of the estimated 100 000 glaciers in the world. While glaciers in most parts of the world have receded in recent years, some glaciers in New Zealand and southern Scandinavia have been advancing. Again, an alternative approach to analysing and extrapolating sparse observations is to model glacier responses to observed temperature changes. Both estimates and observations indicate a reduction of mass of glaciers and ice caps, giving a contribution to global average sea levels of 0.2 to 0.4 mm yr^{-1} over the last 100 years.

Total melting of the Greenland ice sheet would raise sea levels by around 7 m, before the subsequent longer term isostatic adjustment. Precipitation on Greenland is equivalent to around 1.4 mm yr^{-1} of MSL fall. For the Antarctic ice sheet, total melting would raise sea levels by around 61 m and the annual accumulation is equivalent to 5.1 mm yr^{-1} of global sea level fall. This means that small changes in the balance of accumulation on the ice sheets through precipitation, and the loss through ablation (evaporating to the atmosphere) and production of icebergs is a critical factor in the sea level budget. Direct measurements of the volume of these ice sheets are now possible using satellite altimetry, aircraft radar and (more precise) laser altimetry, coupled with GPS. In theory, because the ice caps have only about one thirtieth of the area of the oceans, measuring ice volume is a very attractive way of checking their contribution to global MSL trends: a rise of 10 mm in MSL over the global ocean is equivalent in volume to a 300 mm reduction in the thickness of the polar ice caps. However, satellite coverage at high latitudes is limited (TOPEX/Poseidon goes only to 66° latitude and the ERS satellites only to 81° latitude; see Section 1.4.4).

In looking at the balance of the polar ice caps it is important to remember that there is also a long-term ongoing adjustment to the glacial/inter-glacial transition as the ice sheets themselves are not yet in equilibrium with today's air temperatures.

7.5.3 Other effects

Possible extra contributions to present-day sea level change may be variations in the amount of water removed from the sea and stored elsewhere, either naturally (as in ice caps) or artificially. Storage includes water retention in lakes and reservoirs, in groundwater and as permafrost. These latter hidden redistributions of water mass are difficult to detect, but it may be possible to do so using gravity satellite data. The changing patterns of human water management and land use could be affecting the measured sea level changes. For example, building more reservoirs would to some extent reduce the rise in sea level due to climate warming. The MSL equivalent of water impounded behind reservoirs is 140 mm, a capacity that has increased steadily in recent decades. Conversely, the abstraction and use of groundwater for irrigation could have increased sea levels in the range 0.0 to 0.5 mm yr^{-1} over the past century, with higher values in recent years.

Terrestrial water storage is one of the least certain factors in estimating past and future sea level changes and their causes, but a best estimate for the past century suggests an average collective effect of between -1.1 and $+0.4$ mm yr^{-1}. Permafrost and sedimentation could also make small MSL contributions, as shown in Table 7.1.

7.6 Summary of recent MSL changes

Before looking at possible future MSL changes in Chapter 8, it will be useful to summarise our understanding of the causes of the changes over the past 100 years and longer. There is still considerable uncertainty about the exact contribution of these various physical factors to the observed increases of global MSL over the past century. While there is approximate agreement between the observed changes and the changes computed from our understanding of the physical processes, the observed MSL changes are generally higher than those modelled (see Figure 7.11). There are many uncertainties and the ability of ocean models to simulate decadal changes in ocean temperatures, thermal expansion and circulation is still improving. There is also a great deal of uncertainty in the contribution of the mass balance of the Greenland and Antarctic ice sheets. Finally, the unrepresentative distribution of long-term sea level measurements over the twentieth century and the uncertainty in the correction for land movements means that, despite careful averaging, a true global MSL increase can still only be estimated.

People often ask whether the rate of sea level rise is accelerating. Against the background of natural shorter term variability it has proved very difficult to detect any changes in the *rates* of MSL rise, although these are anticipated in theoretical responses to global warming. There is evidence in the longest records that rates of MSL change increased in the last few decades of the nineteenth century, but careful analysis of twentieth-century records has failed to identify further acceleration. The collection of good data over longer periods and improvements in our ability to make realistic adjustments for effects such as air pressures, winds and ocean currents are necessary before reliable accelerations might become evident.

The Intergovernmental Panel on Climate Change (IPCC), which reports every five years on the most recent scientific evidence, reaches its conclusions after a rigorous process of consultation and debate. The following summary, based on their most recent conclusions, encapsulates many of the issues that have been discussed in this chapter.

- Since the last glacial maximum about 20 000 years ago, sea level has risen by over 120 m at locations far from present and former ice sheets, as a result of loss of mass from these ice sheets. There was a rapid rise between 15 000 and 6000 years ago at an average rate of 10 mm yr^{-1}.
- Based on geological data, global average sea level may have risen at an average rate of about 0.5 mm yr^{-1} over the last 6000 years and at an average of 0.1 to 0.2 mm yr^{-1} over the last 3000 years.
- Vertical land movements, sometimes comparable to rates of eustatic MSL rise, are still occurring today as a result of these large transfers of mass from the ice sheets to the ocean.
- During the last 6000 years global average sea level variations on time scales of a few hundred years and longer are likely to have been less than 0.3 to 0.5 m.
- Based on tide gauge data, the rate of global MSL rise during the twentieth century is in the range 1.0 to 2.0 mm yr^{-1}, with a central value (not necessarily the best estimate) of 1.5 mm yr^{-1} (see Figures 7.8 and 7.11).
- Based on the very few long tide gauge records available, there are indications that the average rate of sea level rise in the nineteenth century was less than that during the twentieth century.
- No significant further acceleration of the rate of sea level rise during the twentieth century has been detected.

These conclusions will become firmer as more data accumulate and as our scientific understanding and modelling techniques improve. In particular, local details will be much clearer when a few decades of data from altimeters and long series of GPS fixing of benchmarks are available.

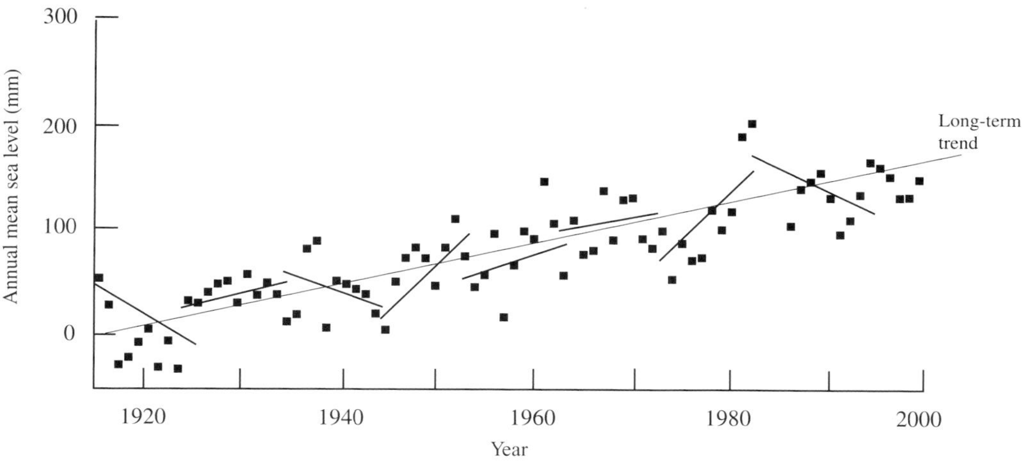

Figure 7.12. Annual MSLs at Newlyn over 85 years. Short lines are fitted to successive ten-year blocks of data, and several show a downward slope. This shows the dangers of estimating long-term trends from data covering only ten years or less.

In the next chapter we discuss how present risks of coastal flooding are estimated and how these might change as global climates alter.

Misleading short-term trends

The year-on-year variability of annual MSL makes the determination of long-term trends from short periods of data very unreliable. This has not prevented enthusiastic but misguided inexpert analysts extrapolating data covering as few as five years over several hundred years to predict changes of many metres. The reliability of trend estimates depends on the length of data, the amount of inter-annual variability, the quality of the data and the reliability of the datum control. Figure 7.12 shows the considerable variations of individual ten-year trends at Newlyn. Against the background of inter-annual variability, it appears from studies of trends worldwide that several decades of annual MSLs are needed at a site to determine the local long-term trend of MSL relative to the tide gauge benchmark. This is confirmed in Figure 7.13, where trends, after correcting for GIA, appear to become more stable for records longer than 50 years; the remaining scatter is probably mainly due to unreliable data and vertical land movements other than GIA.

One scientific challenge is to relate the inter-annual sea level variations worldwide to long records of air pressures and winds collected

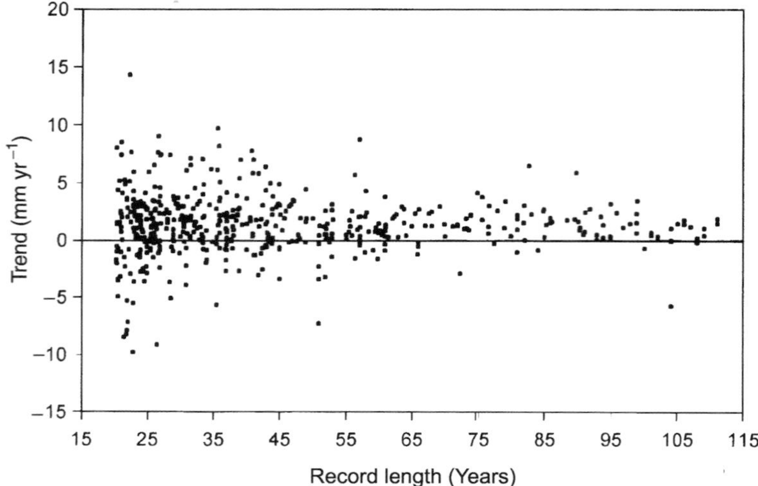

Figure 7.13. This figure shows that sea level records of fewer than 25 years have the greatest variability in MSL trends. The longest records tend to cluster around trends of 1–2 mm yr^{-1}. The values here have been corrected for GIA (from Douglas *et al.* (2001), with permission).

over the past century. If this is done we would learn much about the corresponding ocean circulation, and be able to adjust the tide gauge records of annual mean sea levels; long-term trends would then be much easier to identify.

Further reading

Mean sea level has become and remains an active area of research, stimulated both by popular concern about global warming and by a range of new measuring capabilities such as altimetry and GPS. Woodworth *et al.* (2004) give an excellent account as do Douglas *et al.* (2001). Fu and Cazenave (2001) is a comprehensive description of many aspects of the subject. In the latter two books, the chapters by Nerem and Mitchum are especially relevant. Cabanes *et al.* (2001) have looked at satellite and tide gauge measured sea level changes and related them to ocean thermal expansion. Vertical land movements are discussed by Murray-Wallace (2003) in a way that complements the time scales of our own treatment of sea level changes. The Intergovernmental Panel on Climate Change (Church *et al.* 2001) regularly reviews and updates the evidence. The Intergovernmental Oceanographic Commission manuals on sea level measurement analysis (IOC 1985, 1994, 2001) explain different filtering and averaging techniques for calculating MSL. Readers interested in the addictive pursuit of trends based on old sea level records will enjoy reading Maul and Martin (1993), Woodworth (1999) and Pugh *et al.* (2002). The PSMSL website (as detailed on the website accompanying this book) is a valuable introduction to wider aspects of the subject.

Questions

7.1 How much will MSL be in error if the positive half of a tidal cycle of amplitude 3 m is included in the arithmetic mean for:

(a) a month (59 tidal cycles)
(b) a year (705 tidal cycles)?

7.2 What is the difference between MTL and MSL at a tide gauge where $\mathbf{M_4}$ is a maximum at high water and has an amplitude of 3 per cent of the $\mathbf{M_2}$ 3 m amplitude?

7.3 Calculate the mean current of the Gulf Stream where there is a 1 m level difference over a width of 100 km, at 30°N. Use the value of f in Table 4.4 and assume $g = 9.81\ \mathrm{m\,s^{-1}}$. Why are currents greater than this often observed?

7.4 Explain briefly why the observational evidence is inadequate to separate MSL changes observed in the twentieth century due to ocean temperature changes and those due to melting of grounded ice.

7.5 How will observational evidence used to distinguish the causes of MSL change be improved in the twenty-first century?

Chapter 8
Extreme sea levels

8.1 Return periods and risk

One of the main reasons for studying sea level changes is to predict flooding risks, and especially how these might change in future. Before looking at possible future changes we should look at the general concepts of risks and how these are calculated for extreme sea levels. The first two sections in this chapter may be omitted by readers who do not need to understand the mathematical concepts. The remainder of the chapter looks at observed trends and some of the potential coastal impacts.

Increasingly, coastal planners have to include estimates of flooding risk into their designs, and allow for an appropriate measure of protection against expected extreme sea conditions during the lifetime of any proposed development. Careful assessment of the probabilities of extreme sea levels is a necessary part of the design of modern coastal infrastructure systems. Estimating these risks needs to be based on good data and a range of analysis techniques.

There is a necessary and important distinction between knowing the risks and ignoring them. Known risks can be assessed from observations using probability theory and can be incorporated into planning, investment and defence design; if the risks are not known there is no basis for making decisions, other than 'trusting to luck'.

The probabilities of extreme sea levels and coastal flooding may be specified in several different ways. These levels, including the tide, surge and mean sea level elements, are sometimes called *still water levels* to distinguish them from the total levels, which include waves. Waves are usually accounted for separately in risk analyses, although more elaborate procedures may allow for some correlation between storm surges and high-wave conditions.

If the probability of a level z being exceeded in a single year is $Q(z)$, that level is often said to have a return period, which is in the inverse of $Q(z)$ in years. For example, a sea level having a probability of being exceeded in a year of 0.05 would be said to have a return period of 20 years. Similarly the level that has a probability of being exceeded once in a hundred years is called the *100-year return level*. This inversion of annual exceedance probabilities to give return periods makes the implicit assumption that the same statistics are valid for the whole period specified; since for very small probabilities this may be many tens or hundreds of years, this can be a very big assumption. It would be absurd to say the 10^{-4} level has a 10 000-year return period because MSL conditions would have changed substantially over that period.

The appropriate value of $Q(z)$ chosen for coastal planning will depend on the value of the property at risk. Nuclear power stations may specify 10^{-5} or 10^{-6}. For the coastal protection of the Netherlands a value of 10^{-4} is adopted, but for many British coastal protection schemes values of 10^{-3} or greater are accepted.

One way of presenting the risks of extreme still water levels is as the probability that a stated extreme level will be exceeded at least once during the specified design life of the structure. This is called the *encounter probability* or *design risk*. For the first year the risk is $Q(z)$, and the probability of not reaching that level is $(1 - Q(z))$. The probability of not reaching the level z in either of the first two years is $(1 - Q(z))^2$ and the risk of reaching the level is $[1 - (1 - Q(z))^2]$. As a simple comparison, this is the same as calculating the chance of throwing two dice and not showing a six on either. For a sea level example, if $Q(z)$ is 0.1, the risk of exceeding the level z in the first two years is 0.19. Over several years the *design risk* is related to the annual exceedence probability according to the statistical relationship:

$$\text{Risk} = 1 - (1 - Q(z))^{T_L} \tag{8.1}$$

where T_L is the design lifetime. This is plotted in Figure 8.1 for $Q(z) = 0.01$ (100-year return period). It shows that designing for a level which has an expected lifetime equal to the return period is generally not an acceptable criterion. It should be remembered, as illustrated by the dotted line for the 100-year design level in Figure 8.1, that a structure has a through-life probability of 0.63 of encountering a level that has a return period equal to its design life. For good design, the *design level* must

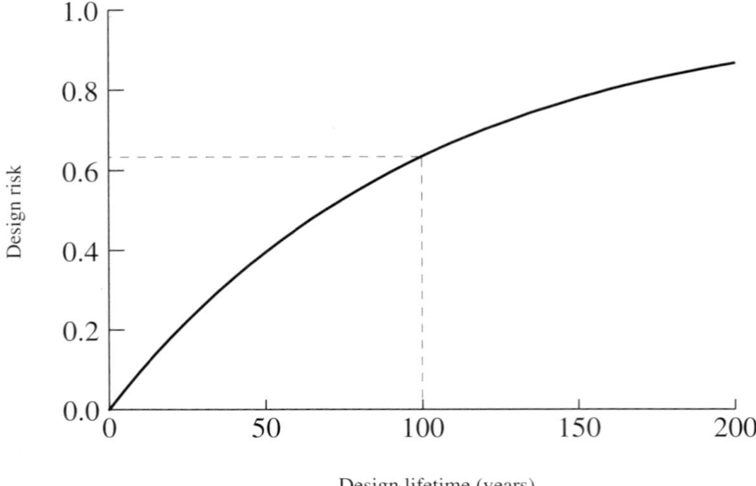

Figure 8.1. The relationship between the risk of encountering an extreme level with a return period of 100 years and the expected lifetime of a structure. For example, the dotted line shows that if the structure is to last for 100 years then there is a 0.63 probability of encountering the hundred-year level during its lifetime.

have a return period that considerably exceeds the expected lifetime of the structure. As an example, suppose that the envisaged life of a structure is 100 years, then for a risk factor of 0.1 of exceedence during this period, the design level should have a return period of 950 years.

When estimating $Q(z)$ it is very unusual for an engineer or coastal scientist to have access to the quantity or quality of data that many of the theoretical techniques described in this chapter require. Estimates must be based on only limited observations at the site proposed, so that extrapolation of the available data in both time and space is inevitable. Skill is necessary to decide how valid it will be to use data from another location, and the best way to make the transfer.

8.2 Ways of estimating flooding risks

Here we outline briefly the ways in which sea level data can be used to estimate risks of flooding. The methods which we will discuss are most effective for calculating extremes for regions outside areas influenced by hurricanes. In these extra-tropical regions extreme sea levels are usually due to a combination of high astronomical tides and extreme weather effects.

For very expensive structures the most elaborate available statistical methods should be used to estimate extremes. For less expensive schemes approximate methods of estimating have been developed as cheaper alternatives. The methods below are applied assuming mean sea level trends have been removed before analysis. For details of the application of the methods, readers should consult specialised publications.

8.2.1 Regional factors

The simplest approach is to compute the ratio between some normal tidal parameter and the level having the specified return period of years (typically 100 years) for a standard port in a region. One such factor is defined as:

$$\alpha_{100} = \frac{\textit{100-year highest sea level}}{\textit{Highest astronomical tide} + \textit{100-year surge level}} \qquad (8.2)$$

The 100-year surge levels may be estimated from a long series of meteorological residuals with suitable extrapolation, or by analytical or numerical models, which relate them to 100-year winds. Clearly α_{100} has a maximum value of 1.0, but this is the most pessimistic case where the 100-year surge level is assumed always to coincide with highest astronomical tide. Building structures for $\alpha_{100} = 1.0$ will almost certainly lead to expensive over-design. In practice the values are lower than this because extreme surges will probably occur with more normal tidal levels (see Section 8.2.3). Around the British Isles α_{100} is typically in the range 0.8 to 0.9, except in the southern North Sea, where the value falls to around 0.72. This reduction is because the local shallow-water dynamics, discussed in Chapter 5, cause large surges to avoid times of high water of astronomical tides; this is a very fortunate interaction as it substantially lowers the levels of potential flooding of London and the Netherlands.

8.2.2 Annual maxima

In order to determine the value of $Q(z)$, the annual exceedence probability at a coastal site, from which return periods and risk factors may be estimated, it is necessary to tabulate the maximum levels reached in each of as many years as possible. Extreme levels have a seasonal cycle (weather effects are generally greatest in winter and the tides are often biggest in March and September) so it would be wrong to use values from periods shorter than a year.

The annual maxima of Newlyn data over 84 years are plotted as a histogram in Figure 8.2. The level of highest astronomical tide (3.0 m) was exceeded in only 28 of those years. The broken curve in Figure 8.2 shows the probability of a particular level being exceeded in any single year. For example, the probability of an annual maximum level exceeding 3.0 m at Newlyn is 0.33, because 28 yearly maxima in the set of 84 were higher than this. Expressed in a different way, an annual maximum in excess of 3.0 m has a return period of three years. Plots like Figure 8.2 are useful for representing the general characteristics of

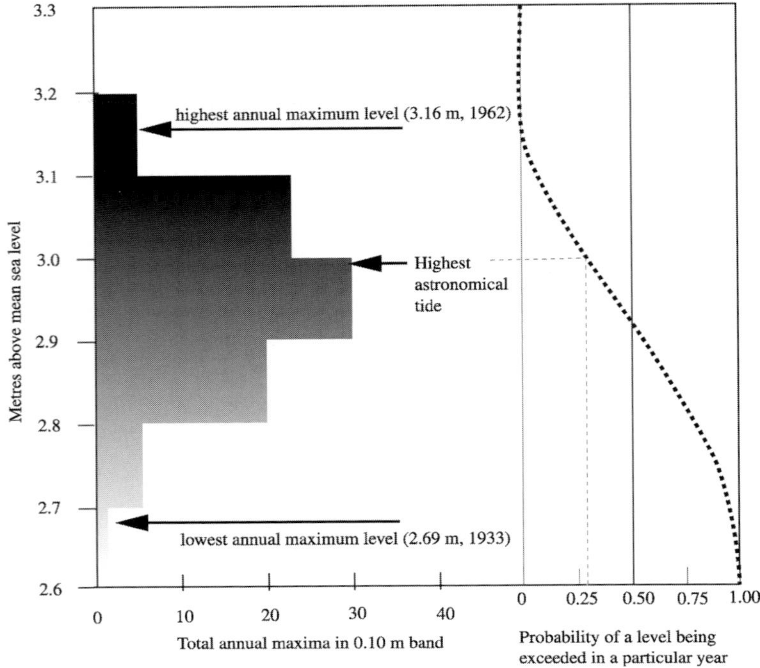

Figure 8.2. Observed annual maximum sea levels at Newlyn, 1916–2000, and the probability of a level being exceeded in a particular year. Mean sea level trends are removed from the data. The most probable annual maximum level lies in the band 2.9 to 3.0 m above mean sea level. The fine dotted line shows that the highest astronomical tide (3.0 m) has a probability of 0.33 of being exceeded in a particular year.

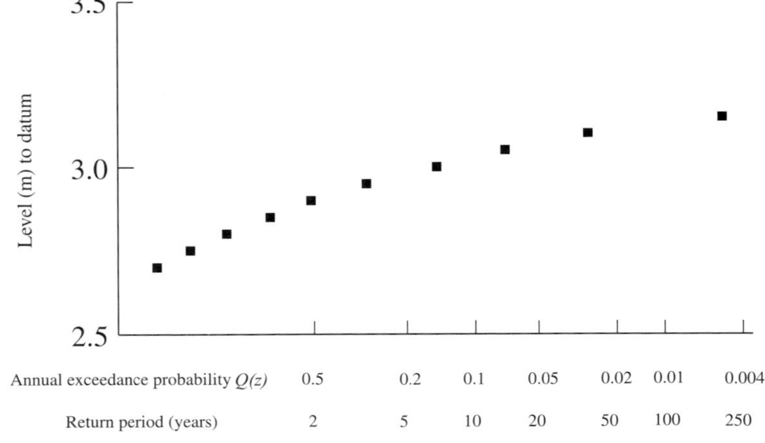

Figure 8.3. A different way of showing the information in Figure 8.2. This shows the probabilities of annual maximum levels at Newlyn falling below a specified level. Mean sea level trends have been removed.

annual maxima, but they cannot be used for the extrapolations necessary when estimating for extreme events which, by definition, have a very low probability and value of $Q(z)$.

The usual procedure is to fit a curve to values of z plotted against the probability of annual exceedence, as in Figure 8.3. Plotting the levels against an x-axis logarithmic scale for probability has the advantages of opening out the two ends of the probability curve ($P = 0$ and $P = 1$) relative to its central position, and of making the transformed curve approximately linear. (For fuller details of the process, consult the publications recommended in the Further reading section at the end of this chapter). The family of curves used for fitting and extrapolating to very low probability events is known as the generalised extreme value (GEV) distribution. The most appropriate curve is usually obtained by the method of least-squares fitting to the data.

Although as few as ten annual maxima have been used to compute probability curves, experience suggests that at least 25 annual values are needed for a satisfactory analysis. As a general rule extrapolation should be limited to return periods not longer than four times the period of annual maximum levels available for analysis, but even within this limit extrapolated values should be interpreted with caution. Experience also shows that the form of the extrapolated curve is almost always strongly controlled by the last few points of the plotted values; it is often found that one or two extreme levels observed during the period appear to lie outside the usual distribution pattern. The degree of weight given to these becomes a matter for subjective judgement. They cannot be discounted easily: the dangers of omitting the most extreme, genuine sea level values from an analysis are obvious.

The major disadvantage of the annual maxima method is the waste of data, because a complete year of observations is being represented by a single value. If the largest meteorological surge for the year coincides with a low tidal level, the information is ignored despite its obvious relevance to the problem of estimating extreme level probabilities (see Section 8.2.4).

8.2.3 Joint tide–surge probability estimates

An alternative way of estimating probabilities of extreme levels is to make use of the separate distribution of tidal and surge frequencies. Tidal probabilities can be determined from quite short periods of data by tidal analysis because the range of tidal forcing is well known from the astronomy. Figure 8.4 shows the statistical distribution of predicted tidal level at Newlyn over an 18.6-year period. The double-humped distribution, with the most frequent levels near to mean high and mean low water on neap tides, is typical of semidiurnal tidal regimes. The frequency distribution of non-tidal (surge) levels is plotted in a similar way in Figure 6.1.

Table 8.1 shows how joint tide–surge probabilities can be calculated in practice. In this example we assume that the extreme levels occur

Table 8.1. *Example of high water and high water residual probabilities for calculating joint probabilities. For example, a surge of 0.1 m represents all surges in the range 0.05 to 0.15 m.*

		Non-tidal residual (m)				
Tidal HW		−0.2	−0.1	0.0	0.1	0.2
level (m)	*Normalised frequency*	*0.1*	*0.2*	*0.4*	*0.2*	*0.1*
3.2	*0.1*	0.01	0.02	0.04	0.02	0.01
3.1	*0.2*	0.02	0.04	0.08	0.04	0.02
3.0	*0.3*	0.03	0.06	0.12	0.06	0.03
2.9	*0.3*	0.03	0.06	0.12	0.06	0.03
2.8	*0.1*	0.01	0.02	0.04	0.02	0.01

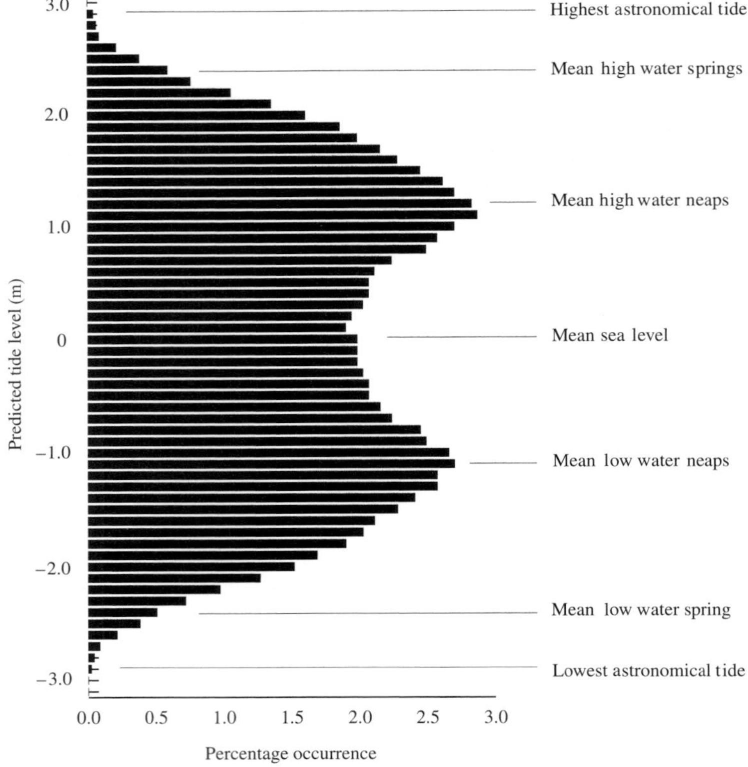

Figure 8.4. The frequency distribution of hourly tidal levels at Newlyn over an 18.6-year nodal period. The intervals are 0.1 m. The most probable levels are at mean high water neaps and mean low water neaps.

on a high tide, when there is also a large positive surge. The residual (surge) and predicted (tidal) high water levels over a complete number of years are tabulated to produce normalised frequency distributions. An appropriate tabulating interval is 0.1 m. In the example shown in Table 8.1 the tidal high water levels and the meteorological distributions have been artificially restricted to five 0.1 m class intervals in each case. Forty per cent of the observed residuals were in the range of −0.05 m to +0.05 m whereas 10 per cent were in the range 0.15 m to 0.25 m. The highest tides lie in the range 3.15 m to 3.25 m above the defined datum, represented by the 3.2 m band.

The frequency distributions of the surge observations and the tidal predictions are then assumed (this is not a trivial assumption) to be representative of the probability of future events. The joint probability of a 3.2 m predicted tide and a 0.0 m surge is 0.04, the product of their individual probabilities. Similarly, a 3.1 m tide and a 0.1 m residual have a joint probability of 0.04. For a 3.0 m tide with a 0.2 m residual the probability is 0.03. Any of these three joint events will produce a total observed high water level of 3.2 m, and so the total probability of a 3.2 m level, obtained by scanning along the diagonal line in Table 8.1, is the sum of the three probabilities, 0.11, i.e. 11 per cent of all observed sea levels will lie between 3.15 m and 3.25 m.

In this example the highest total level, 3.4 m, can only occur when a 3.2 m tide and a 0.2 m residual coincide, which has a joint probability of 0.01. When this method is applied to real data much smaller probabilities of extreme joint events are calculated, because the probabilities are distributed over many more class intervals.

In this case there is a natural way of relating these probabilities to the time interval between tidal high waters. For example, 3.4 m levels occur on average once in every 100 tides. More generally these dimensionless probabilities are converted to return periods using time-scaling factors related to the persistence of extreme events (see the Further reading section).

The principal advantages of the joint tide–surge probability approach may be summarised as follows.

- Stable values are obtained from relatively short periods of data. Even a single year can yield useful results, but four years is desirable, to sample several storms.
- There is no waste of information.
- The probabilities are not based on large extrapolations.
- Estimates of low water level probabilities are also produced.

However, data must be of good quality, with timing accuracy to better than a few minutes. If there are timing errors (old chart recorders were especially prone to these) tidal variations will appear in the non-tidal residuals. Joint tide–surge probability estimates of extremes require a high degree of analytical skill; extra computational effort is also involved.

8.2.4 Other methods

Other methods, either similar to or extensions of the annual maxima and joint probability methods described briefly above, are often applied. One method looks at a fixed number of the maximum extreme levels in a year, typically up to ten, instead of just the single annual maximum value. Sometimes this will mean including some highest levels from one year, which are less than those not included for another year. Care is necessary to make sure that each storm is a separate and independent event, which is sometimes taken to mean that they occur at least three days apart. These storm levels are then fitted with an appropriate extreme value distribution.

Another method, developed for river hydrology, is to look at the number of times a level exceeds some stated threshold level. Unlike the previous method, this may result in several storms being recorded in one year and few or none in another year. Total sea levels (tides plus weather effects) could be influenced by the 18.6-year cycle in tidal amplitudes (Figure 3.2). A better approach is to apply this peaks-over-threshold method to the non-tidal residuals, after removing the tides, and to use the resulting probabilities in a joint probability calculation.

The reliability of all estimates is limited by the available data and by possible trends and changes in the regional meteorology and oceanography. The possibility of some rare unsampled event, such as a tsunami, cannot be ignored nor is it easily incorporated into the estimates. Tsunamis are more common in the Pacific than in other oceans, but even for the Atlantic, well-documented tsunamis have occurred (Section 6.7). On the west coast of the United States and in Japan, tsunamis are recognised as the most important cause of extreme levels.

In tropical areas, for example, the Atlantic and Gulf Coast of the United States, extreme sea levels are produced by hurricanes; however, these are too rare at any particular place to permit the calculation of reliable probabilities from observations. The tidal contribution is usually a much smaller factor than the weather (see Section 6.5). A modelling approach, where all possible hurricane characteristics and tracks are simulated, is often used to estimate the very small probabilities.

8.3 Risks and climate change

Popular interest in increased risks of coastal flooding in future warmer climates usually concentrates on the effects of MSL changes. However, as we know from previous chapters, MSL is just one of the three factors, with tides and weather, that affect total observed sea levels. In this section we look at past trends and possible future changes in all three, and how these changes might influence future total flooding risks.

8.3.1 Tidal changes

Here we can summarise the information given in the box (Are tides changing?) in Chapter 4. The major part of the day-to-day variability of coastal sea levels is normally due to tides. Any change in the tidal 'constants' for a site will lead to significant changes in the probabilities of extreme total sea levels, and so must be considered in making long-term projections. Coastal tides are the result of propagation of tidal energy from the deep oceans, where the tides are generated by gravitational tidal forces. Secular changes in the astronomical tidal forcing are known to be negligible in our time scale of interest, because they must be caused by secular changes in the dynamics of the sun–earth– moon system, which is extremely stable on the century time scales of interest here.

Ocean tidal amplitudes also depend on the response of the oceans to these tidal forces and on the local responses of shallower seas. Again, the shape and depth of the oceans and seas will not change much over the time scales of interest here. If MSL increases, tidal wavelengths will increase and tidal patterns will be stretched. However, except in very shallow water the expected changes will be small.

Nevertheless, as already discussed in Chapter 4, locally there have been significant changes in tidal amplitudes. There is good evidence from long-term tide gauge records that tidal ranges can vary locally over decades and centuries (see also Section 9.7). Particularly in estuaries, local siltation, changes in dredging practices for navigation and canalisation of rivers are all factors that influence tidal ranges. Changes in shelf tides due to increasing MSL and the corresponding altered propagation characteristics of tidal waves are much smaller.

8.3.2 Trends in weather effects

Future changes in the frequency and intensity of storms will affect the probability of coastal flooding. So too may changes in the pattern of storm tracks because of the complicated way in which the shelf seas respond to weather systems. These changes are very difficult to anticipate and interpretation is often controversial. Looking for trends in observed extreme events is not a reliable analytical tool: one problem is that data on extremes are by definition scarce, and become increasingly rare for the most extreme events of interest. An alternative is to simulate events using numerical models, but here too there are difficulties; environmental responses become increasingly non-linear and difficult to model as the

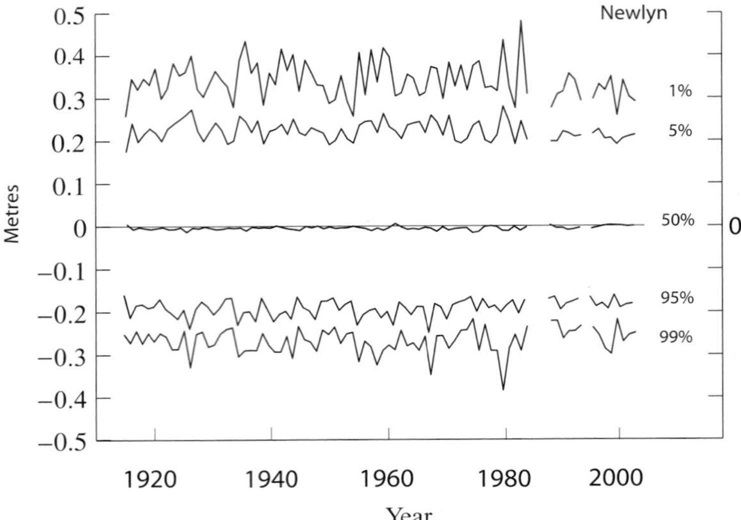

Figure 8.5. Percentile levels of surges at Newlyn, 1916–2000. Annual mean sea levels have been removed in the analysis. These curves show the surge levels, with the stated percentage of time for which they are exceeded (with acknowledgement to Isabel Goncalves Araújo).

energy in forcing becomes more extreme. On a global scale the outputs from global climate models do give indications of changes in storminess, but at present these are in an early stage of development.

There are many ways of analysing sea level data for trends in non-tidal weather effects. Figure 8.5 shows the statistics of weather effects at Newlyn over the period 1916–2000. For each year the levels exceeded 99, 95, 50, 5 and 1 per cent of the time are plotted. Although there is substantial variation from year to year, especially for the very extreme (99 and 1 per cent) levels, there are no discernible trends. An alternative approach where few reliable sea level observations are available is to run computer models of the sea level response to air pressure and winds, and to analyse the computed sea levels for trends. This is possible because the air pressure and wind datasets are relatively long and complete.

Ocean–atmosphere computer models incorporating changes due to the expected increases in greenhouse gases allow tentative estimates of storminess in future climates. In the northern hemisphere, for double the present level of atmospheric carbon dioxide, mid-latitude storm track activity may be shifted northwards and intensified. This modelled effect is particularly marked for the Atlantic storm track and northwest Europe, but is also evident for the North Pacific. The effects of these increases would be to increase the probability of sea level extremes due to surges on the coasts of western Europe, the western United States and Canada. When these changing storm patterns are applied to the waters of shallower shelf seas, changes of up of to 0.2 m in the five-year extremes have been modelled in the English Channel. However, similar models of the North Sea showed no difference in the frequency of extreme surge levels. Increases in water depths due to rising sea levels would tend to reduce wind effects as discussed in Section 6.4.2.

Possible future changes in the frequency and intensity of tropical storms and their effects on flooding are also difficult to quantify. Tropical storms are related to numerous factors, most notably sea surface temperature. If there is a general greenhouse warming effect, sea surface temperatures are expected to rise with a consequent increase in tropical storms and risks of coastal flooding. The geographical distribution and impact of these changes is an active topic of research.

8.3.3 Expected MSL changes

Past changes in MSL have been discussed in Chapter 7. In anticipation of climate change, likely future trends of MSL have been studied intensively; the Intergovernmental Panel on Climate Change (IPCC) brings together the results of the latest research at five-year intervals. There are two basic approaches to estimating future MSL: the extrapolation of existing observations and the numerical modelling of the earth's response (especially ice and ocean responses) to a warmer atmosphere.

As discussed in Chapter 7 (see Figure 7.2), natural inter-annual variability in sea levels, probably related to changes in ocean circulation, means that detecting reliable long-term trends needs upwards of 20 years of data, and 50 years is better. Underlying trends can be seen more easily if other effects such as the inverted barometer effect and local winds are removed. The inter-annual variability has characteristic periods (for example El Niño over 4–5 years), which vary at different locations. Air pressures substantially affect annual MSL at Newlyn in the northeast Atlantic, but there is only limited correlation in the MSL record at periods over two years. Conversely, the annual MSL at Sydney, Australia is more strongly related to local winds and has a correlation over periods of five years and longer, with negative correlation after 30 years. Each site is different and requires individual analytical treatment.

The search for 'accelerations' in sea level rise continues (Section 7.6). There is no evidence for any acceleration in sea level rise in data from the twentieth century alone, but the few very long records from before the start of the twentieth century (Amsterdam, Brest, Liverpool, Sheerness, Stockholm) suggest that an acceleration occurred during the latter part of the nineteenth century after a period of relatively stable sea levels. Comparison of the rate of rise over the last 100 years with the rate over the last two millennia from geological evidence also implies a comparatively recent acceleration.

The second method is more scientific and versatile. Sea level changes are modelled numerically for a range of global ocean atmosphere models.

These models variously account for increasing carbon dioxide and aerosol emissions, and ice dynamics, as well as for the absorption of heat into the ocean and its transfer to all depths. Section 7.5.1 explains why the levels at which the heat is absorbed in the ocean are important in determining the actual volume expansion that results.

According to the IPCC, summarising outputs from a range of models and emission scenarios, the changes between 1990 and 2100 could be in the range 0.11 to 0.77 m. The main contributions to this rise are expected to be:

- thermal expansion of 0.11 to 0.43 m, accelerating through the twenty-first century;
- a glacier contribution of 0.01 to 0.23 m;
- a Greenland contribution of −0.02 to 0.09 m;
- an Antarctic contribution of −0.17 to 0.02 m.

The estimated total range includes smaller contributions from the thawing of permafrost, sedimentation and ongoing effects of ice sheet melting following the last glacial maximum. Using a wider range of emission scenarios the IPCC project a sea level rise of 0.12 to 0.88 m from 1990 to 2100, with a central value of 0.48 m. These values can be compared with Figure 7.11 and are much greater than the increase through the twentieth century summarised in Section 7.6.

One of the most difficult things to forecast is the actual timing of increases; these are delayed by the thermal inertia of the oceans and ice sheets, and the slow transfer of heat to all depths of the oceans. Even if greenhouse gas concentrations were stabilised at present levels, MSL would continue to rise for hundreds of years before reaching equilibrium. As already explained, some of the present rise in MSL is probably due to atmospheric temperature increases over hundreds of years past, and even to residual effects from the last glacial maximum. For the next centuries the observed MSL changes will be largely due to delayed responses to past temperature changes in the atmosphere and oceans, not to the immediate and contemporary temperature changes.

The west Antarctic ice sheet, which contains enough ice to raise global MSL by 6 m, has been thought by some scientists to have the potential for rapid melting, though general agreement suggests this is not likely to accelerate in the twenty-first century. This effect is not included in the above estimated ranges for MSL rise.

Global models are not yet very consistent in their prediction of the regional distribution of sea level change, although they do show that the range of regional variations will be substantial compared with the overall global average sea level rise. Figure 8.6 shows calculated MSL changes worldwide from one of the models. Nearly all models project a lower

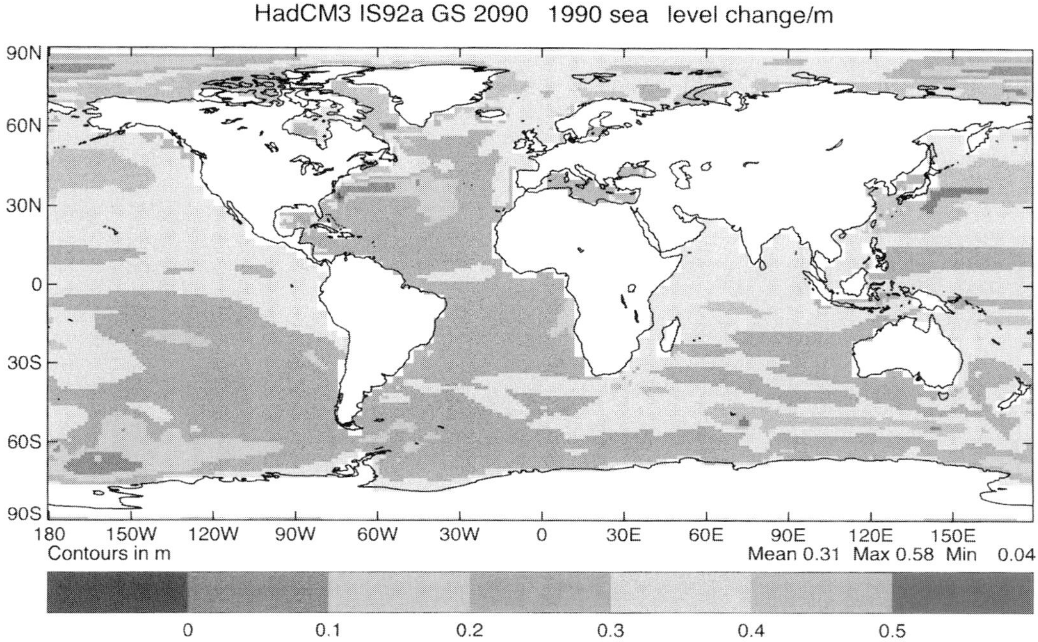

Figure 8.6. Computed sea level changes in metres between 1990 and 2090 for particular emission scenario. This map is based on the results from a coupled global–ocean atmosphere model and supplied by the UK Meteorological Office Hadley Centre. There is not yet strong agreement among the various models, except to show lower than average MSL rise in the Southern Ocean.

than average rise in the Southern Ocean, where, as explained, the colder waters there have a low coefficient of thermal expansion.

8.3.4 Combined effects on flooding risks

The possible changes in the three individual components of observed sea levels can be combined to estimate future risks of various total sea levels. These revised risks can also include expected local vertical land movements and, to be specific, may be related to some future date or to changing risks over the design life of a coastal defence structure. The new plots will have the same *form* as Figure 8.3, but the values will be changed over time.

Figure 8.7 shows the effect of just increasing MSL on risks of coastal flooding. For simplicity we assume a straight-line relationship, with values close to those for Newlyn but without the flattening curve shown in Figure 8.3. The solid line represents the result of an analysis of data from 1916 to 2000. The dashed line shows the effect of raising MSL by

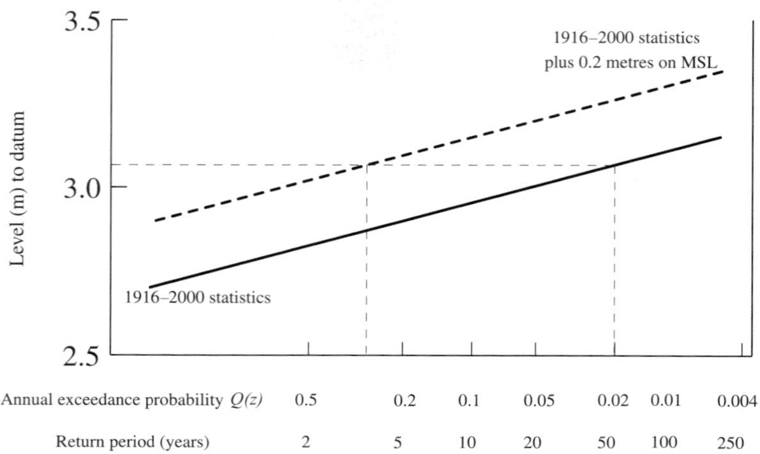

Figure 8.7. The effects of a MSL rise on flooding risks. The solid line is a simplified version of the curve plotted in Figure 8.3. The heavy dashed line adjusts the 1916–2000 statistics for a 0.2 m rise in MSL. The fine dashed line shows that a level of 3.1 m at Newlyn, now reached once every 50 years, will then be reached approximately every four years.

0.2 m, but no other change. In the former case, the 50-year return level is 3.1 m; after the rise in MSL this level has a return period of about four years.

The shape of the curve relating extreme levels to return period is critically important when considering sensitivities to future changes. The effect of rising sea level depends on the shape of the curve, which is different for each individual port. Figure 8.3 shows a flattening off for the more extreme events, which is often but not invariably the case. If the curve is flatter, the effect of rising sea level on reducing return period is even more dramatic than at Newlyn. Vulnerable places that have flatter curves, such as the Pacific Islands, will experience the greatest change in flooding risks as MSL rises.

One uncertainty in estimating changes in flooding probabilities is the way in which the shape of the curve may change, for example through increased storminess in addition to rising MSL. Calculations of combined effects for places around the United Kingdom have suggested changes in extreme levels that differ from adding only a MSL increase, as we did in the previous paragraph, by as much as ±0.20 m.

We must end this section on assessing flooding risks with a word of warning. However elaborate the computations of extreme sea levels, the result should always be treated with caution because of the limited period of observational data compared with the equivalent return periods being calculated. The possibility of some rare event cannot be ignored, nor is it easily incorporated into the estimates. As a particular example, tsunamis, which are generated by seismic activities and are rare events, are unlikely to be represented in the statistics of the observed data (see Section 8.2.4). Similarly, where extreme sea levels are produced by hurricanes, these are usually too rare at any particular place to permit the calculation of reliable probabilities from observations. In these cases the modelling and simulation techniques described in Chapter 6 are more appropriate.

8.4 Responses to changing flooding risks

8.4.1 Impacts of changing sea levels

A sensation-hungry press often dramatises possible MSL rise by showing maps of changed coastlines for rises of 5 m or more. The *expected* changes over the next hundred years of around 0.5 m discussed in the previous section are an order of magnitude less. The potential impacts are much more modest, but in some cases will still be serious unless preventative measures are planned and implemented. In all discussions of the impacts of an increase in MSL, we must remember that the *rate* of change is probably more important than the *magnitude*. If the rate is slow enough, then natural, social and economic systems will adapt at their own rates.

The following paragraphs give brief descriptions of some of the major potential impacts, as summarised by the IPCC and in other sources listed in the Further reading section at the end of this chapter.

Flooding

Increased risks of coastal flooding are almost certain for most coastal areas. Even places which now have falling relative MSL because of Glacial Isostatic Adjustment, may see the fall reversed if increased rates of MSL rise take effect. Flooding has both immediate (Figure 8.8) and longer term consequences. The initial impacts in terms of danger to life and disruption of activities are obvious. In the longer term, there could be damage to infrastructure, which includes roads and railways, energy supply networks and resort amenities. Once agricultural land has been flooded with saline water it takes a long time to recover.

All of these have economic impacts. In many places there will also be dangers of disease because of contaminated water supply and damaged septic systems. Urban flooding will be a problem where water supplies, drainage and waste management systems can be overwhelmed. In some cases modern drainage systems and hard-constructed surfaces give more rapid discharge of flood waters, which can make things worse if high river levels and high sea levels coincide. Removal of groundwater for water supply can also increase risks of flooding due to the resulting related local subsidence (see Figure 7.10).

According to the IPCC there are ten million people today who live in coastal areas with a flooding return period of a year or less. Increased

Figure 8.8. A typical *aqua alta* scene in St Mark's Square, Venice. The trestle walkways on the right are assembled when high sea levels are predicted (photograph copyright City of Venice, with permission).

coastal populations and rising MSL could raise this to as much as 240 million people by 2080. The areas at risk are found around the globe, but particularly vulnerable areas will be west Africa, east Africa, the southern Mediterranean, south Asia and southeast Asia. Some of the biggest flooding impacts are on small island states with no capacity for the population to move to higher areas inland.

Coastal erosion

About one third of the world's coasts consist of sand or shingle beaches, and there is evidence that most of these have been eroding in recent years. However, there are many factors other than MSL changes which may be responsible for the observed erosion, including loss of sediment supply due to construction of defences elsewhere and changing patterns of beach use. For a change in MSL, the response of a beach profile may be estimated by the widely used Bruun rule of coastal erosion as illustrated in Figure 8.9. By equating sediment volumes in the erosion and deposition zones it may be shown that the shore erosion resulting from an increase of MSL ΔZ_0 is given by:

$$\text{Shoreline recession} = \left(\frac{\text{Active beach width}}{\text{Offshore depth}}\right) \times \Delta Z_0 \qquad (8.3)$$

At its simplest this merely states that the average beach slope will be maintained as MSL rises and there is a corresponding inland movement

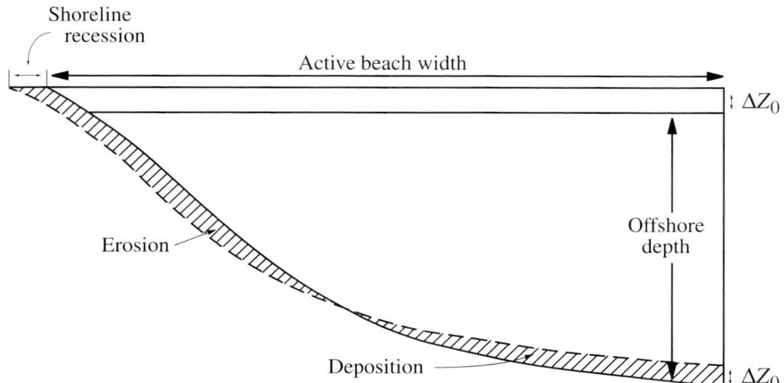

Figure 8.9. The Bruun rule of coastal erosion (Equation (8.3)) is widely used by engineers to predict the effects of changing sea levels.

of the shoreline, but the result also applies for more complicated beach profiles. This rule is developed on the assumption of a closed sediment system; a closed system means that there is no loss or additional supply of sediment except from the beach and near shore, over the active beach width. Despite the limiting two-dimensional assumptions made in developing this formula, it has been applied with some success for beaches along the east coast of the United States, particularly when allowances were made for vegetative matter in the eroded material and for known along-shore sediment transport.

The Bruun rule does not suggest that MSL rise directly causes erosion. Coastline movement is the result of the waves attacking further up the beach and transporting sand offshore. Even where there are coastal cliffs, if these are formed of soft materials they can be eroded progressively as the waves reach, attack and undermine their base more frequently.

Coral reefs

Coral reefs are found in a zone within 30° of latitude north and south of the equator. Many low-lying islands such as the Maldives consist of, and are protected by, coral reef structures. Reefs are built by coral polyps, small animals that extract calcium carbonate from the seawater and grow progressively into solid rock formations. These polyps need water above 18°C to grow, but above 28°C their growth is inhibited. They cannot endure long exposure to the atmosphere nor to the extreme temperatures that may result from direct exposure to sunlight. They grow best in near-surface layers where there is good light penetration. Upward growth of the reef may not be able to keep up with a rapid MSL rise, and the resulting reduced light levels will slow the coral growth. The maximum rates of upward coral growth depend on many factors including the type of coral, but rates between 5 mm yr^{-1} and 10 mm yr^{-1} are often quoted.

Reef growth may also be affected by other climate changes such as upwelling and colder water temperatures. In the Red Sea movement of the zones of wind convergence has recently caused enhanced upwelling and algal growth, which in turn has tended to smother the coral and inhibit growth. Higher water temperatures (and possibly other factors) in the Indian Ocean during 1997 led to a general bleaching of corals. Other possible climate-related factors that can affect reef growth include wave activity and insensitive tourism.

Coastal wetlands and salt marshes

Coastal wetlands and salt marshes, which develop at levels near the high tide water level (see Figure 9.4), are particularly vulnerable to MSL rise. Many are found at the upper reaches of estuaries. Today, after a relatively stable period of sea level, extensive mangrove ecosystems are common. They act as sediment traps to stabilise sedimentary coastlines, provide protection against hurricanes and storm surges, often serve as nurseries for commercially exploited crustacean and fish species and are a natural resource base for wood and a large range of economic products. The continuing supply of sediments and organic material is critical. Studies suggest that sediment accumulation generally allows wetlands to build up at rates of up to perhaps 2 mm yr^{-1} to 3 mm yr^{-1} depending on the availability of sediment. However, they are in danger of being overwhelmed if the sea level rise is faster than accumulation rates. If this happens, natural coastal defences will be lost. One estimate quoted by the IPCC suggests that as much as 22 per cent of the world's wetlands may disappear by 2080.

For the Louisiana coast of the United States, the former accretion process has now reversed to an erosion regime. Canalisation by dredging of rivers can be a significant factor in reducing sediment supply because the higher speeds of tidal currents carry river sediments directly out to sea. The Mississippi delta now loses about 100 m a year as a result of reduction in sediment input from the river and sea, the intrusion of saltwater, and deterioration of vegetation. Interior brackish ponds are formed that eventually lead to the root death of marsh plants. Although salt-tolerant species may be able to replace the freshwater species, this is not always possible, and wetlands may convert to shallow lakes. Inland extension of wetlands is possible, with the majority of plants and animals shifting gradually upwards, but only if the adjacent land is low lying and available. If the existing coastal limits are constrained by the construction of defence barriers, inland extension is not possible.

Saltwater intrusion

Extraction of water from the upper reaches of estuaries for domestic use and for irrigation would be affected by sea level rise. The effects can be compared with the salinity increases at times of drought and low river flow. Drainage of low-lying areas by opening sluice gates at low tide levels would become less effective as the time for drainage is reduced. A study of the Ijssel Lake in the Netherlands, which is below mean sea level, has shown that natural drainage would remain possible at low tide for a MSL increase of 0.5 m. Further increases, however, would require expensive pumping stations and higher protecting dykes for the surrounding land.

8.4.2 Is it worth paying for protection?

Usually, yes. One estimate suggests that by 2100, 600 million people will live on coastal flood plains below the 1000-year flood level. The number of people at risk from annual flooding as a result of a 0.4 m sea level rise and population increase is expected to increase from today's level of ten million to 50–80 million by the 2050s and 240 million by 2080. In 2050 more than 70 per cent of these people will be concentrated in a few regions mentioned earlier: west Africa, east Africa, the southern Mediterranean and the south and southeast of Asia. Small island states are particularly vulnerable to all the impacts listed in Section 8.4.1. Estimating the economic costs of potential damage, and even of protective measures, is exceedingly difficult because there are so many uncertainties involved. When choosing strategies for defence, planners balance the risks, the value of amenities and the cost of protection.

The value of coastal amenities is very high, as approximate global estimates published by the intergovernmental Organisation for Economic Cooperation and Development illustrate. A sea level rise of 1 m by 2100 (at the high end of the IPCC forecast range) would cost $970 billion worldwide if fully protected against. For the United States alone, estimates suggest a loss of 1500 homes per year and a cumulative impact on coastal property by 2100 ranging from $20 to $150 billion. However, the cost will be substantially more than this if coastal resources are either over- or under-protected. Over-protection means that too much money will be spent on defences; under-protection means that the value of the coastal amenity will be lost. In each local area at risk, appropriate cost–benefit analysis is necessary.

The problem of responding to projected sea level rises has global, regional and local implications. At one extreme, vast resources might be expended on protecting the present coastal position and facilities. Unless protective measures are designed with full understanding of the

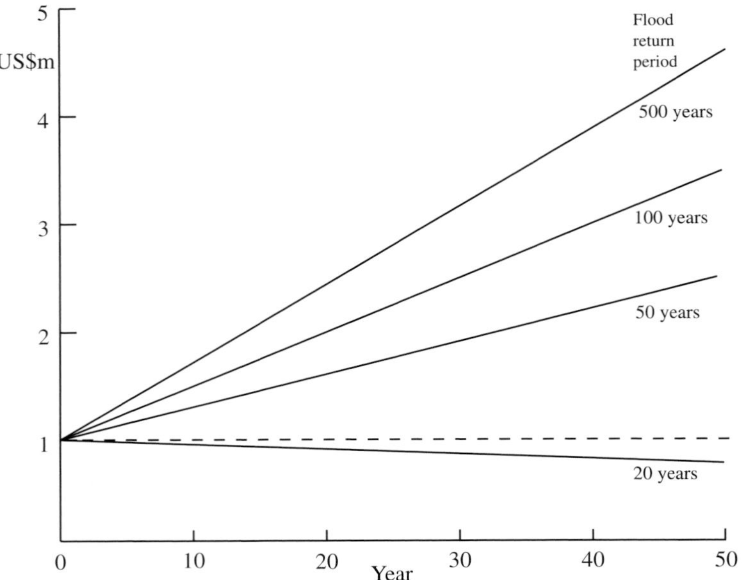

Figure 8.10. This graph shows the statistical return from a US$ 1 million coastal asset, assuming a US$ 500,000 protection investment in year zero, repaid over 50 years. Income is 10 per cent of asset value per annum and interest rates 5 per cent of the outstanding loan. Income is not re-invested and there is no allowance for inflation. A flood event is assumed to cause a total loss of the asset. Protection for a 20-year return level is a poor investment, but higher levels of protection are worth considering.

implications, they could result in negative impacts such as increased erosion elsewhere. The complete defence solution would be very difficult to apply for areas such as the Ganges and Nile deltas, and for low-lying coral islands such as the Maldives and many others in the Pacific. For the first metre or so of MSL rise defences are probably quite feasible and cost-effective for many of the tidal cities of developed countries, for example, London and Rotterdam. At the other extreme, a minimum response would be to plan an ordered withdrawal to higher land. The gradual adjustment of low-lying land prices to reflect their finite availability will be the financial response over many decades, similar to the present purchase of leasehold property for a fixed period. For these kinds of economic adjustments to work effectively the market will need good estimates of the changing risks.

Cost–benefit analyses are the basis of these assessments. The *costs* of defence will include capital for building a barrier or other form of protection to reduce flooding risk, payment of interest on the capital and any subsequent maintenance costs. The *benefits* will include continuing use of existing amenities and the generation of profits from their use, protection of the original value of the assets and, perhaps, reduced compensation payment for public liabilities. The future values of all of these factors are much more uncertain even than the effects of climate change. So, inevitably, are the calculated cost–benefit assessments.

Figure 8.10 shows, as an example, the accumulated value of a defended coastal asset for different risks, annual profitability and interest rates. It shows the accumulated value of a US$ 1m coastal asset, assuming US$ 0.5m is invested in protection in year zero and repaid over 50 years. Value increase is 10 per cent of the asset per annum and there are interest repayments at 5 per cent of the outstanding loan. Income is not re-invested and there is no allowance for inflation. A flood event is assumed to cause total loss of the asset. Protection only for a 20-year return event is a poor investment, but investment for higher levels of protection are worth considering. These calculations also show that the uncertainties in non-environmental economic factors such as interest and profitability are at least as important as the changing risks of flooding in controlling the likely return on the investment.

If the future economic conditions such as interest rates and profits are uncertain, why bother with measuring risks of flooding? Most analyses show that over a long period risk reduction by investing now in defence systems can usually be justified. In the long term, coastal land values have traditionally increased in real terms as population pressures grow, and will probably do so in the future. Nevertheless, for some low-amenity land the most cost-effective option may be a managed retreat, despite popular local protests.

Public opinion should not be ignored, whatever the results of cost–benefit analyses. Financial planning is only a part of the overall political and social decision-making process, and all these factors will influence the final decision. Only local people can decide what value they place on their assets and amenities and how much they will pay to protect them. Estimating cash values is particularly difficult for environmental amenities, which have traditionally not been given an economic value in a strict market sense. What value do present generations place on recreational beaches, everglades and wetlands? And how can we take into account their value for future generations?

The international Framework Convention on Climate Change incorporates agreements among governments on the progressive reduction of carbon dioxide emissions. Nevertheless, there will continue to be increases in atmospheric carbon dioxide and other greenhouse gases for several more decades. Global temperatures will continue to rise. So too will sea levels and the risks of coastal flooding. Finally we must remember that ocean volumes have not yet fully adjusted to past increases of global temperature.

8.4.3 Examples of responses

Carefully planned adaptation is called for. The purpose of adaptation is to reduce the net cost of climate change and sea level rise, whether those costs apply to an economic sector, to an ecosystem or to a country.

Planners have identified three response strategies to increasing sea levels: protect, accommodate or retreat, as summarised in Figure 8.11. In the past few years there has been an increasing emphasis on managed coastal retreat, where this is possible. Managed retreat aims to avoid hazards and to prevent ecosystems being squeezed between artificial developments (such as defence walls) and the advancing sea. This can be effected by legislation so that new buildings must be set back at some minimum distance from the shore, or by denial of flood insurance.

Although sea level change is only one factor in the dynamics of coastal engineering and management, coastal regions should urgently assess their present risks of flooding and then look to the future. For estimating future risks there are many uncertainties that can be reduced by research, but there is no substitute for maintaining good local sea level measurements.

Many countries that have vulnerable low-lying coastal areas recognise their responsibilities. Several are now developing national and local monitoring and response programmes. Cooperative programmes of measurement and analysis are good value and make best use of scarce data and limited resources. Here we describe briefly a few examples of local and national responses.

Several Caribbean countries have worked together under funding from the World Bank and the Global Environment Facility to improve their monitoring of sea level. They have developed databases and information systems, an inventory of coastal resources, and will monitor sensitive coral reefs and assess coastal vulnerability and risk. The coordination of twelve countries working together within a programme is administratively challenging, but the advantage of exchanging sea level data and analysis skills far outweighs the potential initial difficulties.

In the United Kingdom, the Government has issued guidelines for sea defences that incorporate enhanced MSL rise. The recommended factors have been adapted to take into account the different post-glacial vertical land movements in the north and south of the country. Allowing for the land movement, a medium–high risk probability design level for the west of Scotland by 2050 incorporates an increase of 0.17 m, whereas along the sinking coast of East Anglia the design increase is 0.37 m. Movements of tide gauge benchmarks are being regularly monitored by GPS, as described in Section 7.4. A new network of GPS and tide gauges has been installed to monitor movements around London, which is already protected by the Thames Barrier. The policy message is that it will probably be economically optimal for Britain to protect most but not all of its vulnerable stretches of coastline. A small area of agricultural

Current Sea Level

RETREAT | ACCOMMODATE | PROTECT

Buildings

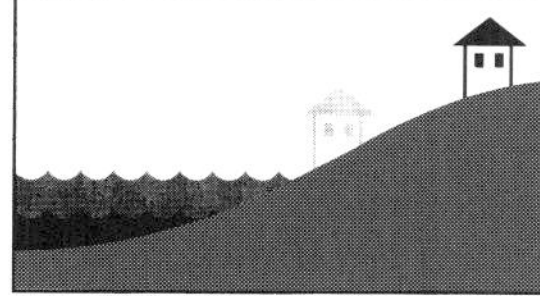

Establish building setback codes

Regulate building development

Protect coastal development

Wetlands

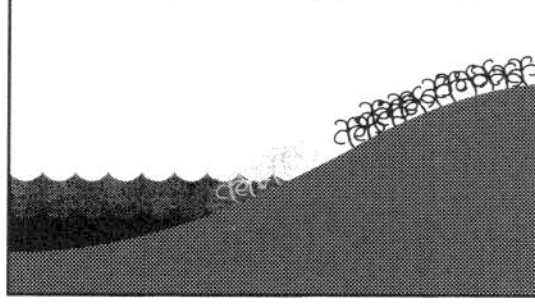

Allow wetland migration

Strike balance between preservation and development

Create wetland/mangrove habitat by landfilling and planting

Crops

Relocate agricultural production

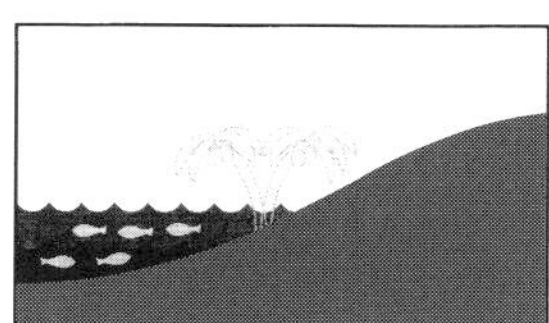

Switch to aquaculture

Protect agricultural land

Figure 8.11. Different response strategies to rising sea levels. The strategies fall into three categories: retreat, accommodate or protect. All three options are not always possible. For example, retreat is not possible on low-lying tropical islands (from IPCC, with permission of Cambridge University Press).

Figure 8.12. Managed retreat of the coastline is sometimes the preferred option rather than defence against erosion. Here at Tollesbury, Essex on the east coast of England, sea defences are being breached deliberately to form new coastal wetlands (photograph copyright English Nature).

land at Tollesbury (Figure 8.12) on the East Anglia coast of the North Sea was re-opened to flooding from the sea in 2002 to create new salt marshes, as part of a local managed retreat strategy.

In the Netherlands, computer studies of possible changes in the extreme wind climate of the southern part of the North Sea have shown that for this particular area the changes are likely to be small in comparison with other areas. In Spain, three complementary tide gauge networks, operated by different public bodies, have been integrated into a new system. Similar programmes and studies are in place in many other countries.

Around the coast of South Africa, which touches three oceans, a measurement programme has collected sea level data from ten sites and related the risks of flooding to the value of the amenities that are at risk. As in many other countries, the sea level data are often only short term and it will be necessary to re-evaluate the risks as new data become available. Here, by carefully selecting the key sites for measuring sea level, some interpolation to estimate risks at intermediate locations will be possible. Australia has set up an annual sea level reporting indicator of change, as part of a national assessment process. This will have other uses: for example, MSL at Fremantle, Australia is used to predict lobster recruitment onto coastal reefs and subsequently into the local lobster fishery.

In Canada, areas of detailed local study include St John, New Brunswick; Charlottetown, Prince Edward Island; and several other communities on the Atlantic coast. At Charlottetown, Prince Edward Island, surveys of sea levels and low-lying areas have shown $US 80 million of property at risk, even for a flood level already reached in January 2000 (see Figure 6.9). By 2050 that flood level will have a probability of 0.8 of being reached at least once every year, without increases in the present rate of MSL rise.

In the United States the areas most at risk are the mid-Atlantic and south Atlantic States, because of their low-lying land, high economic value and frequency of storms. The coastal islands of southern New England are also vulnerable. Along the mid-Atlantic east coast the barrier islands, separated by lagoons and connected by occasional bridges to the mainland, are especially vulnerable; here there are already vital evacuation procedures in place for hurricanes and flooding. Homeowners are expected to evacuate their homes by following defined exit routes if a hurricane is approaching and are not allowed to return until the emergency is over.

On the west coast the risks are generally lower, but important exceptions include San Francisco Bay and Puget Sound. The total coastal land area in the United States that would be inundated by a 0.5 m rise in MSL is around 24 000 km^2. To avoid the damaging consequences of flooding, several coastal cities – including New York, Washington DC, Miami and New Orleans – will need to improve their defences.

Small low-lying ocean coral islands, the Small Island Developing States (SIDS), are especially vulnerable as there is nowhere to go if a tropical storm is approaching. Their problems in planning for future MSL increases include having very limited data, little or no local expertise to assess the dangers, and often a low level of economic activity to pay for the remedial measures. The Maldives consists of more than 1200 islands, none of which rise to more than 3 m above MSL. A 1 m MSL rise would inundate 85 per cent of the capital island, Malé, home to 56 000 people.

In assessing possible impacts of increased flooding risks, many countries have used the systematic 'Seven Steps of the Common Methodology' recommended by the IPCC (see the box at the end of this chapter). This common approach has been defined because it allows comparable worldwide assessments and can help international development and financial institutions to include MSL rise into their decision-making about future investments. The Common Methodology considers potential impacts on population, economics, ecology, social assets and agricultural production. The standard approach also allows and stimulates communication and exchange of data, methods and technology at all levels: political, technical and scientific. The challenge of rising MSL is a global issue that calls for global cooperation.

Seven Steps of the Common Methodology

Seven Steps have been recommended by the Intergovernmental Panel on Climate Change for assessing the vulnerability of coastal areas to sea level rise. Vulnerability includes the nation's ability to cope with consequences of accelerated MSL rise.

Step 1 Delineate case study area; specify accelerated sea level rise and climate change conditions.
Step 2 Make inventory of study area characteristics.
Step 3 Identify relevant development factors.
Step 4 Assess physical changes and natural system responses.
Step 5 Formulate response strategies identifying potential costs and benefits.
Step 6 Assess the vulnerability profile and interpret results.
Step 7 Identify future needs and develop a plan of action.

Further reading

There are several textbooks on the statistics of extreme values: one of the most recent and readable is Coles (2001). For a detailed account of applying different techniques in the United Kingdom, see Dixon and Tawn (1994); there are other specialised publications for other regions. The series of reports issued by the Meteorological Research Centre, Bangladesh, gives a well-researched account of flooding risks in the countries of the northern Indian Ocean. Details of the modelling of extreme sea levels and further references to trends in the three components of observed sea levels are given in Pugh and Maul (1999) and Flather (2000). The most comprehensive accounts of the impacts of MSL rise both globally and regionally are undoubtedly provided in the IPCC publications, 2001 (IPCC, 2001) being the latest at the time of writing. Readers will find that Bird (1993) and Douglas *et al.* (2001) provide briefer, more approachable introductions, as do Woodworth *et al.* (2004).

Questions

8.1 Use Equation (8.1) to show that for a design life of 100 years and a risk factor of 0.1, the design level should have a return period of 950 years.

8.2 If two dice are rolled, what is the chance that both will show a six; that neither will show a six; and that only one will show a six?

8.3 Using similar reasoning, if a sea level has an annual exceedence probability of 0.1 (ten-year return period), what is the chance that this level will be exceeded in both 2007 and 2008; in neither of these years; and in just one of the years?

8.4 What is the approximate probability of the tidal level at any one time, predicted for Newlyn, being more than 2.0 m above MSL?

8.5 From Figure 8.7, what is the new return period for the present 250-year level, after MSL increases by 0.2 m, assuming no changes in either the storm or tidal statistics?

8.6 If MSL rises by 0.5 m in a sea 50 m deep, by how much will an $\mathbf{M_2}$ amphidrome a quarter-wavelength from the reflecting coast move?

Chapter 9
Tidal influences

Although this is the final chapter of our book on sea level changes, it may also be the beginning of a wide range of further studies of their importance and influence. In previous chapters we have looked at the physical reasons for sea level change, particularly change due to tides, weather effects and climate. Here we introduce some of the coastal and biological processes that are controlled or strongly influenced by sea level, particularly by tides. We start with sediment processes by looking at how coastal lagoons and connecting inlets are affected by tides. We then look at the influences of tides on salt marshes and mangrove swamps. The importance of tidal patterns for plants and animals, especially those that have adapted to live on the sea shore between high and low water levels, is also described. We look at how tides and mean sea levels in earlier geological times may have been very different from those we measure today. Finally, we review the use of tidal levels in the legal definition of national and coastal boundaries. These various topics are all specialised scientific areas that can be investigated in much more detail by following the references given in the Further reading section at the end of the chapter.

9.1 Tidal inlets

Around the world many coasts have bays, estuaries or lagoons connected to the ocean. Often this connection is through a narrow inlet channel: the tides inside the bay are controlled by the tidal rise and fall of the level in the ocean and by the geometry of the inlet. Because of the importance of these inlets for water exchange and for navigation, coastal

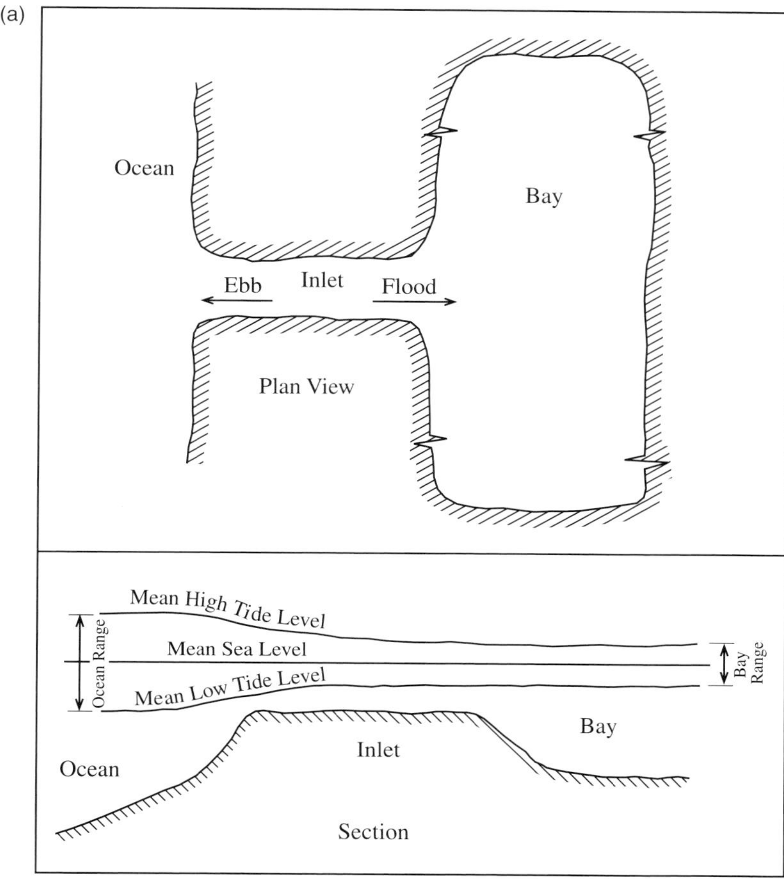

Figure 9.1. (a) Summary of the geographic and tidal conditions for a bay connected to the sea through a narrow entrance. (b) The corresponding ocean and bay tidal levels for St Lucie Inlet, Florida. Bay levels lag ocean levels by more than an hour and the slack water (i.e. no currents; inlet current is indicated by the dashed line) occurs nearly three hours after high tide in the ocean. With acknowledgement to Lee Harris, Florida Institute of Technology.

engineers have studied their development and stability in detail. Figure 9.1a summarises the essential parameters of a bay-inlet system. The volume of water exchanged between the sea and the bay during a tidal cycle is called the *tidal prism*.

9.1.1 Flow in tidal channels

The flow through the inlet is controlled by its cross-section and by the resistance of bed friction. It takes a finite amount of time for the water to flow from the ocean to the bay, which means that slack water (zero tidal

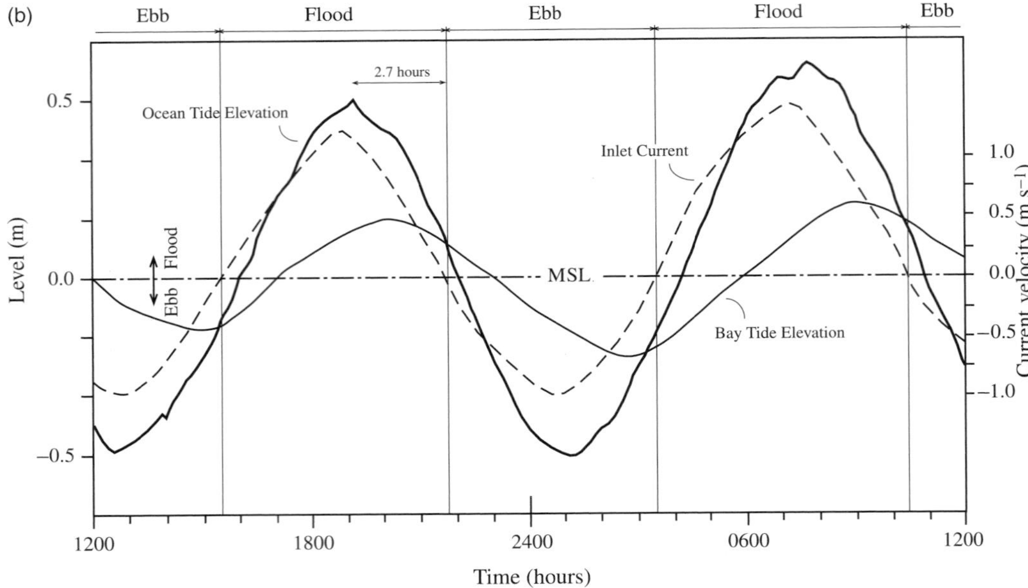

Figure 9.1. *(cont.)*

currents) in the inlet occurs typically two or three hours after high or low water in the ocean. This is shown in Figure 9.1b where the bay tide has a smaller amplitude and lags the ocean tide by about three hours. The flow through the inlet is driven by the sea level difference between the two ends. This is the same as for East River, New York (Section 5.4 and Figure 5.7) although in our present example the tidal variations at either end of the channel are hydrodynamically dependent.

Slack water occurs in the inlet when the sea levels are the same in the ocean and the bay, usually around the same time as high or low water level in the bay. This is typically two or three hours after high water in the ocean for most small inlets. This time lag is sometimes called the slack water phase lag or tidal current phase lag for the inlet. Peak flood flow from the sea into the lagoon occurs very close to the time of ocean high water.

The ebb flow is a maximum close to low tide at the ocean end of the inlet when the inlet water depth is lowest. This means that the cross-sectional area of the inlet will be smaller during ebb flow than flood flow so that for shallow inlets, for the same quantity of water to leave the inlet as enters during a tidal cycle, the maximum ebb flow is greater than the maximum flood flow. During ebb flow the outgoing water in the inlet meets the incoming waves; this together with the greater outflow velocity and the shallowness at the entrance combine to create the most hazardous conditions for inlet navigation during the tidal cycle.

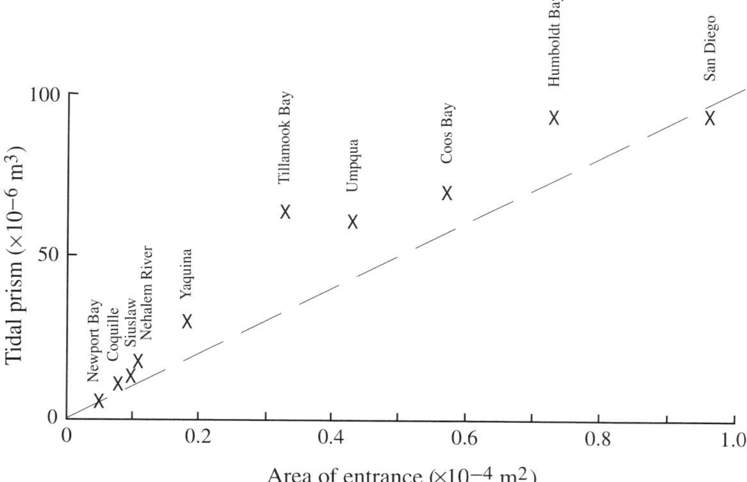

Figure 9.2. The relationship between the cross-sectional area of a bay entrance and the volume of water exchanged (the tidal prism) in each tidal cycle for a series of bays on the west coast of the United States. The dashed line shows the relationship if the tidal prism is 10^4 times the area entrance, both in metric units.

9.1.2 Inlet cross-section and the tidal prism

Figure 9.2 shows the relationship between the tidal prism and the cross-sectional area of the entrance channel at its narrowest point for bays along the Pacific coast of the United States. Along this coast the tides are mixed, and so the tidal prism has been calculated as the volume exchanged between mean higher high water and mean lower low water. There is a close relationship between the volume of water exchanged (the tidal prism) and the minimum inlet cross-section; this statistical relationship holds well for a wide range of coasts and bays worldwide.

This relationship has been examined in detail by coastal engineers as it can be used to predict some of the effects of artificially widening or deepening an entrance channel, or the effects of recovering low-lying tidal flats within a bay. The relationship implies that if land is reclaimed, then there will be a reduction in the entrance area because of a smaller tidal prism. In extreme cases a smaller volume of water exchanged through the channel, for example on neap tides, can lead to it becoming blocked by sediment. A good example of this is the Batticaloa Lagoon on the east coast of Sri Lanka, where access to the sea from the lagoon and fish-processing facilities is periodically interrupted by the closure of the lagoon entrance during small neap tides; when this closure occurs, boats can go fishing only after the fishermen have re-opened the access channel. The general relationship, as illustrated in Figure 9.2 has been found to fit:

$$\text{Minimum cross-section} \times 10^3 = 0.0656 \times \text{tidal prism (in metric units)} \quad (9.1)$$

A more approximate statement is to multiply the cross-section in square metres by 10^4 to get the tidal prism in cubic metres. Equation (9.1)

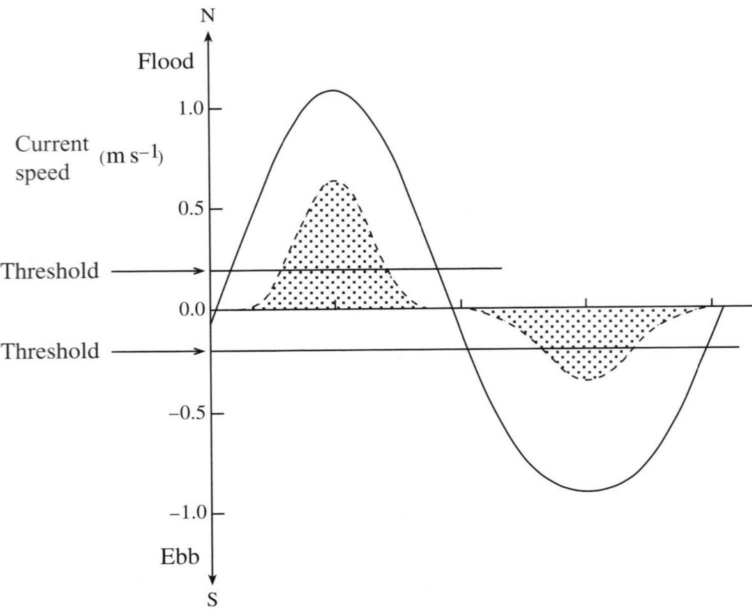

Figure 9.3. The relationship between the net bed-load transport and the asymmetry in the strength of flood and ebb flows, assuming a threshold speed of 0.2 m s^{-1} for sediment movement and a cubic law of transport related to current speed.

has been refined by many researchers to improve its prediction skill by taking into account local sediment supply, river inputs and bay and inlet geometries.

9.2 Tidal asymmetry and sediment movements

Tidal currents flowing into and out of estuaries and other coastal confined areas such as barrier reefs energise a constantly changing regime of sediment flux and coastal dynamics. In Chapter 5 we saw how a tidal wave approaching shallow water steepens so the rise time is shorter than the fall (Figure 5.1). This means that the incoming current (flood tide) is generally stronger than the outgoing current (ebb tide). The sediment transported along the sea bed by currents, called the *bed-load*, increases rapidly as the current speed increases, which means that more sediment is carried inshore than is exported.

We can look at this relationship in more detail. Laboratory experiments show that there is a critical current speed, below which sea bed sediments are not moved. Once the currents exceed a critical threshold speed necessary to get the sediments into suspension, the bed-load transport is approximately proportional to the cube of the current speed; for example, doubling the current speed increases the sediment transported along the sea bed by a factor of eight. Any asymmetry between flood and ebb tidal current speeds can control the rate and direction of net sediment transport. This effect is illustrated in Figure 9.3, where the solid curve shows the current speed (in and then out) and the cubic law related to the speed, indicating sediment transported once the critical speed, here 0.2 m s^{-1}, has been exceeded. The inward sediment flux is twice the outward flux, even though the averaged water flux in and out of the bay is zero.

For tidal flow in a channel the amplitude of the $\mathbf{M_2}$ current is normally the dominant factor for controlling the bed-load transport, but the asymmetry, which decides the direction of net sediment flux, depends on the phase difference between the $\mathbf{M_2}$ and $\mathbf{M_4}$ currents (see Section 5.3). If there is a mean current flow added to the tidal oscillations, this can enhance net sediment transport in its direction of flow.

Formulae developed for steady channel flow conditions in the laboratory may require modification when applied to oscillating tidal flows in the sea. For example, sediments raised into suspension at times of maximum currents will be transported by the flow at later times as they settle back to the sea bed. Also, turbulence and the associated bottom drag, which moves the sediments, are greater on the decelerating phase of a tidal cycle and less when the currents are increasing. There is an increasing delay for tidal current reversals at higher levels above the sea bed, and this can have important implications for suspended sediment transport. This lag can significantly alter the directions and volumes of sediment transport.

In general, the laws that show sediment transport increasing rapidly at higher current speeds mean that sediment transport will be most significant in restricted locations and at times of storms. This is when extreme currents are generated by tide and surge currents acting together, and when simultaneous wave activity helps to initiate erosion.

9.3 Salt marshes and mangroves

Although salt marshes (Figure 9.4) and mangrove swamps appear very different, they both consist of salt-tolerant plants. Both have been established by these plants slowing water movements so that suspended sands and muds settle out and are not subsequently eroded. As a result, there is a progressive increase in the levels of both salt marshes and mangrove swamps over a period of years. A sandy or muddy inter-tidal area is first colonised by species that are most tolerant of salt and submersion; these plants then begin the process of trapping sediments. On many sandy beaches Spartina grasses are the first to become established at the upper tidal levels. As the sediments accumulate, other less tolerant but more competitive species begin to dominate as the shore levels continue to rise. A network of creeks develops for tidal inflow and outflow.

Mangrove swamps support rich ecosystems, comprising terrestrial species in the branches of the mangroves and marine species within the

Figure 9.4. A typical saltmarsh development, below sand dunes on the east coast of England (photograph copyright English Nature).

underlying roots and sediments. Mobile animals such as crabs are particularly well adapted to such conditions. At low tide, tree-living snakes and monkeys are able to feed on the crabs. In addition to mobile organisms, there is also a rich surface-living epifauna attached to tree trunks and branches. This includes many species such as barnacles, which are also characteristic of rocky coasts where there is similar strong vertical zonation of species.

9.4 Zonation of coastal plants and animals

Plants and animals have certain essential requirements if they are to survive and flourish. Each organism develops special characteristics to enable it to compete successfully in its particular environment. At the extremes, conditions on land and in water and the species that live there are very different. At the coast, certain species and ecosystems have developed to thrive in an environment that changes between these two extremes in a pattern defined by the rise and fall of the tide. This habitat is also often subjected to wave action and to strong currents.

For survival in this highly variable region, species must be able to cope with the relatively uniform conditions of submersion: even temperatures, plentiful supply of dissolved oxygen and abundant nutrients, organic debris and micro-organisms that can be extracted relatively easily for food. They must also survive varying periods of exposure to air, causing extremes of temperature, salinity and solar desiccation. Between high and low water tide they may also experience large pressure changes. As a result of these varying degrees of adaptation, a complicated pattern of species zonation can be observed from the bottom to the top of the tidal range.

9.4.1 Patterns of exposure and submersion

Marine biologists have looked for relationships between the local tidal regime and the biological zonation in terms of critical tidal levels (CTLs). There are many different ways of doing this from the basic tidal constituents and tidal predictions, as we now discuss.

Emersion/submersion curves

The extreme difference experienced by a species between emersion (exposure to the air) and submersion by the sea may be presented statistically as in Figure 9.5. This shows the frequency distribution of tidal levels for a year at Newlyn and at San Francisco and the overall percentage of time for which each level is exposed to the air. These frequency distributions (see also Figure 8.4 and Figure 8.5) define the levels at which disturbances and stress due to wave activity are most likely to concentrate. The percentage exposure plots are called exposure or emersion curves; it is easy to present the same statistics in the form of submersion or immersion curves.

Diurnal and semidiurnal emersion patterns

One of the critical factors for any particular inter-tidal plant or animal is likely to be the length of the periods of exposure to air that it must survive. Figure 9.6b, which shows the tides for a particular day at San Francisco, may be used to illustrate the existence of five different zones. These are separated by higher high water (HHW), lower high water (LHW), higher low water (HLW) and lower low water (LLW). For the first zone, above

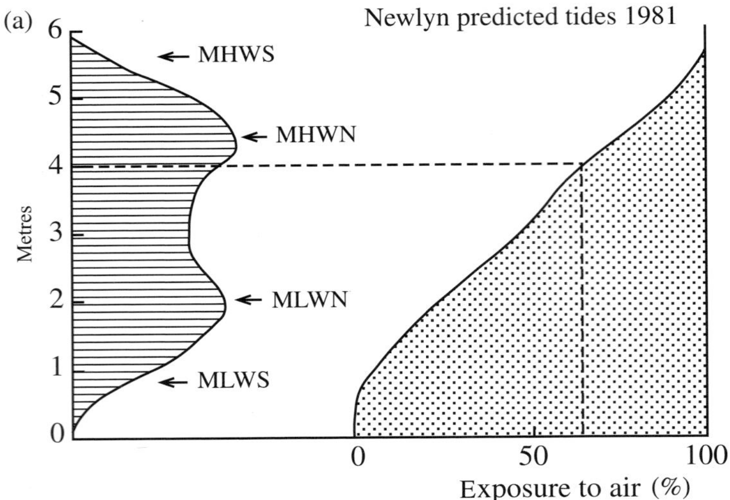

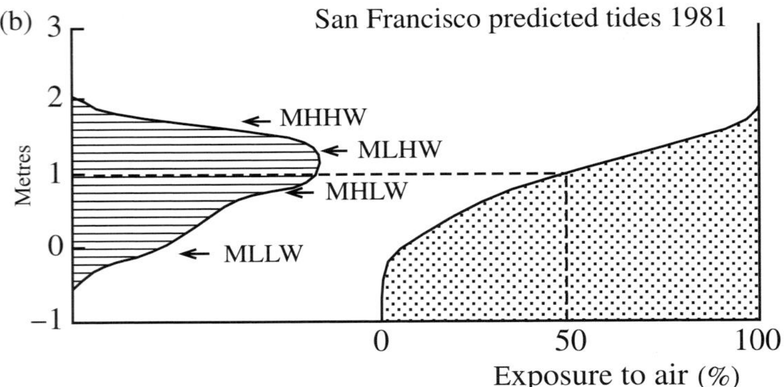

Figure 9.5. The frequency of various tidal levels at (a) Newlyn (semidiurnal tides) and (b) San Francisco (mixed tides) for a year of data. The Newlyn frequencies are shown in more detail in Figure 8.4. The curves on the right-hand side show the fraction of time each level is exposed to air or submerged. At Newlyn the dashed line shows that a level of 4 m above chart datum is exposed to the air for approximately 70 per cent of the time. At San Francisco a level of 1 m above chart datum is exposed to the air for approximately 50 per cent of the time. Key: MHWS, mean high water springs; MHWN, mean high water neaps; MLWN, mean low water neaps; MLWS, mean low water springs; MHHW, mean higher high water; MLHW, mean lower high water; MHLW, mean higher low water; MLLW, mean lower low water.

HHW there is continuous emersion (exposure to air). Between HHW and LHW flooding occurs only once during the day and there is a long period of emersion. Between LHW and HLW there are two submersion and two emersion events in the day whereas between HLW and LLW there is a single relatively brief exposure period. Below LLW there is continuous submersion. Figure 9.6c shows the substantial changes in these zones over a month at San Francisco. At Newlyn (Figure 9.6a), because diurnal tides are small, zones 2 and 4 are very narrow.

Emersion period frequency distribution

A more complete picture of the emersion pattern at any particular level is given by calculating for how long a level is exposed. The most regular cycle of emersion and submersion is found at the mean sea level. A plant or animal at a 4.4 m level above chart datum at Newlyn is normally

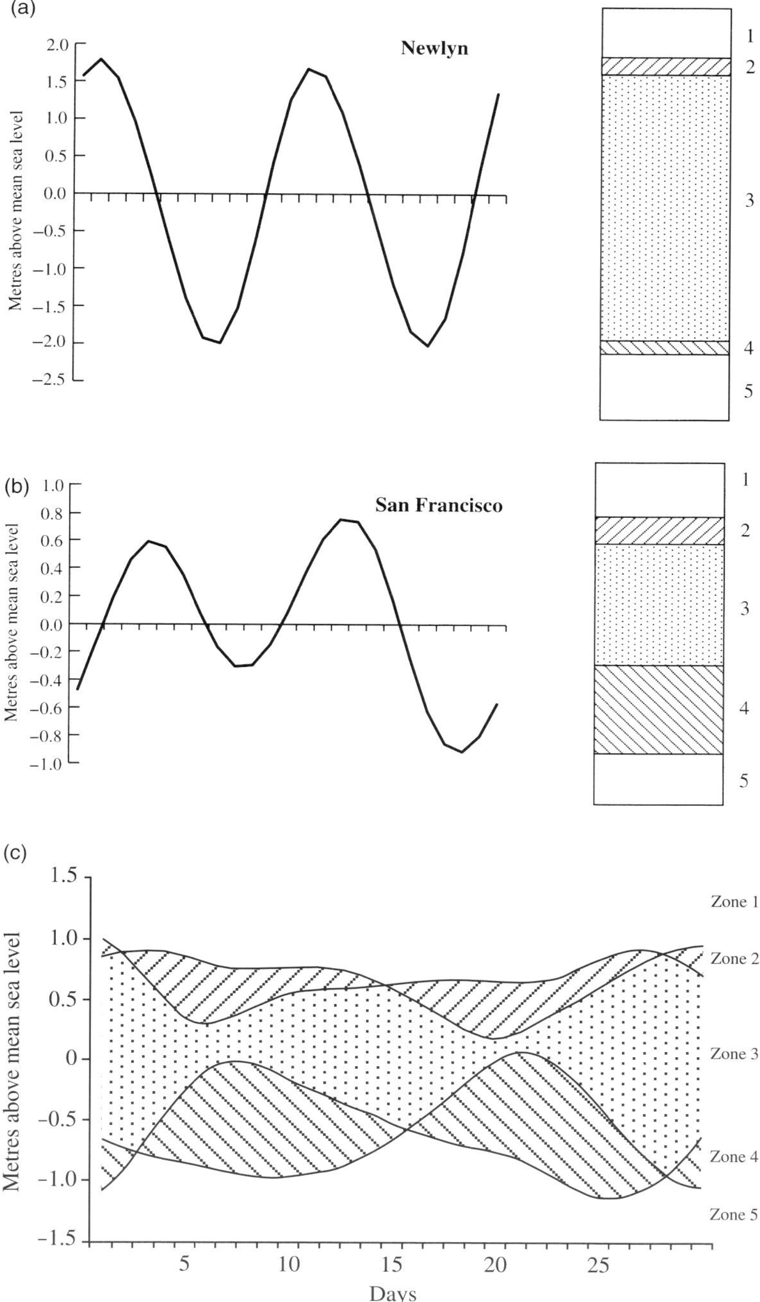

Figure 9.6. Tidal levels at (a) Newlyn and (b) San Francisco for 12 March 2002. The right-hand columns distinguish five separate zones, as discussed in the text: 1, not flooded in the day; 2, flooded once for a short period; 3, flooded and exposed to air twice a day; 4, exposed to air once per day; 5 continuously flooded throughout the day. (c) The variation through March 2002 of the extent of the five zones described in Figure 9.6b for San Francisco.

exposed for periods between 8 and 12 hours but on thirteen occasions in one year it was exposed for periods longer than 30 hours. The longest period of exposure was 112 hours. For San Francisco there are many more periods of exposure of a day or longer because of the greater diurnal inequality.

Time of day at emersion

The dangers of desiccation for a marine plant or animal exposed to the air are obviously much greater if the exposure takes place during the maximum heat of the day. Night-time exposures to winter frosts can also inflict serious stress on plants and animals whose normal preferred habitat is within the more stable temperatures of coastal waters. Coral reefs exposed to cold or rainy weather may suffer extensive damage and subsequent decay of cell tissue. For a semidiurnal tide, the time of day at which extreme low water levels occur is related to the phase of the $\mathbf{S_2}$ constituent, expressed in local time. This is a convenient way of first looking at the vulnerability of a coral reef to solar heating.

9.4.2 Rocky shores

These theoretical tidal statistics can be related to the behaviour of plants and animals in the inter-tidal area. The inter-tidal area between highest and lowest water levels generally shows high biological productivity and contains a very rich and diverse range of species. The many reasons for this high level of productivity include the regular availability of nutrients during each tidal cycle in water that is shallow enough for photosynthesis. But, as already emphasised, it is also a region of great environmental stress. The potential for high productivity can only be realised by those species that have highly adapted survival mechanisms. Different coastal conditions require different survival mechanisms. On sands and muddy shores species can burrow to avoid desiccation, but on rocky shores species must develop more robust defences.

The three non-biological factors that determine the distribution of shore species are the patterns of tidal emersion, the degree of exposure to waves and the nature of the bottom material or sub-stratum.

Rocky shores are both exposed and accessible; plants and animals live on the rocks and have no place to hide. Because they are so accessible, more is known about their biology than about any other marine habitat. Such environments show strong horizontal bands or zonations (Figure 9.7a). The highest levels on rocky shores are almost always exposed to air. Here encrusting black lichens and blue–green algae exist alongside periwinkles and some primitive insects. At lower levels there is a succession of zoned periwinkles, barnacles and mussels. These inter-tidal species often develop in intense clusters to create microhabitats

Figure 9.7. Tidal conditions strongly influence biological productivity and the distribution of species. (a) Zonation on a rocky shore community, Shetland Islands, Scotland (copyright Alexander Mustard). (b) Plankton are carried by tidal currents. In channels they provide abundant feeding for fish and coral. This photograph is from Sinai, Egypt, in the Red Sea (copyright Alexander Mustard).

in which more moisture is retained during exposure. Seaweeds develop at the lower levels. Zonation is caused by a variety of tidal and biological interactions. The upper level of a zone is often determined by physical factors; the lower limit is usually determined by biological factors.

The indigenous Manhousat People of Vancouver Island, Canada had a strong dependence on tides for access to their food along the shore. Food from the very low tidal levels was said to be *kwisaap'alhshitl*, 'food which because it is only occasionally eaten, tastes especially good'.

Both waves and surges blur the boundaries of critical tidal levels where these are identified, and this shading of boundaries makes it unproductive to speculate on the influences of a fine structure within the theoretical statistics of CTLs. The sharp boundaries found in the vertical zonation of species must be accentuated by competitive mechanisms within the ecosystems themselves.

There is a general enhancement of growth in rocky channels (Figure 9.7b) through which tidal currents flood and ebb. This is due to the continuous stream of nutrients and essential foods, and because the water there is usually clear of sediments, which encourages photosynthesis. The relative absence of settling sediments in the water also reduces the risk of clogging of the filter-feeding mechanisms of organisms.

Rock pools in the inter-tidal regions also provide special biological niches. Individual species have different physical and physiological mechanisms for coping with changes of exposure, temperature, higher light intensity, salinity and other variables such as pH and the partial pressures of oxygen and carbon dioxide. Many species shelter in small shallow inter-tidal rock pools during periods of low tide, preferring the possible extremes of salinity, temperature and oxygen depletion to total exposure to the air.

9.4.3 Sedimentary shores

Sedimentary shores – coasts formed by gravels, sand and mud deposits – are in a state of continuous erosion and deposition, and their structure is strongly influenced by tides and waves. In many cases the biological flora and fauna are important agents in the dynamics of coastal stability and change. Sedimentary shores that support life vary from very sheltered bays where fine mud accretes to exposed beaches that are re-worked by waves during every storm. There is a strong relationship between the steepness of the beach and the number of species found. Gently sloping beaches have many more species than steeper beaches where the wave energy is more concentrated and destructive.

Zonation is less obvious on sedimentary shores than on rocky shores because most of the species there survive by burying themselves from extremes of emersion, temperature and salinity. Within the sediments large populations of superficially invisible individual species can thrive. When the tide is out the sediments retain a high proportion of water, so that animals burrowing in the trapped water avoid exposure to low salinities. In addition, the sands filter the water during each tidal cycle and concentrate the particulate matter to the advantage of burrowing animals. Birds are often seen feeding at the water's edge as the advancing and retreating tide re-works the sediments and exposes food. Shingle beaches, consisting of gravels and pebbles, are generally too mobile to allow organisms to establish themselves.

Figure 9.8. Many species adapt to tidal rhythms. In Panama the Sergeant-Major fish spawns in the week preceding new and full moon. This photograph was taken five days before Easter 2003 (see the box at the end of this chapter). The male fish is following the female and fertilising the eggs she lays (copyright Alexander Mustard).

9.5 Behaviour adaptation

The Fiddler Crab (*Uca pugnax* or Atlantic marsh fiddler), found along the east coast of the United States from Massachusetts to Florida, provides an interesting example of complicated ecosystem interactions and species adaptation mechanisms. They forage the scum of nutrients left by the retreating tide, so helping to recycle minerals and organic matter; as the tide rises they are able to seal their burrows against flooding. Their reproductive cycle is linked to the tides: mating takes place during spring tides and the female remains in the burrow for the two-week mating period, before the eggs are released and swept to sea on the next spring tide.

The mummichog, an estuarine species of fish found along the east coast of North America from Florida to the Gulf of St Lawrence, has a spawning pattern related to tides. The eggs are laid at levels reached only on spring tides, hidden in leaves or empty mussel shells. Hatching takes place after two weeks when the eggs are again reached by the tides.

On the Californian coast the grunion, a small fish, comes out of the water to lay eggs a few centimetres deep in the wet beach sand, for three or four nights after the spring tides. Spawning occurs primarily at night from March through to August. Eggs hatch after about ten days, when they are released from the sand by the working of the next series of high spring tides. The baby grunion hatch 2–3 minutes later and are then washed out to sea. Figure 9.8 shows another example of species adaptation to tides.

The movements of plaice, which spawn in the Southern Bight of the North Sea, are adapted to the patterns of tidal currents. In the late autumn maturing fish migrating from more northerly feeding grounds into the Southern Bight are caught most frequently by mid-water trawls on the south-going tidal currents. In winter, spent females returning to the north are caught in mid-water more often on the north-going tidal currents than on the south-going tidal currents.

Clearly, travelling in mid-water when the tidal currents are in the direction of intended migration allows a fish to travel the maximum distance for the minimum expenditure of energy. This so-called selective tidal stream transport has also been noted in sole, cod, dogfish and silver eels. The mechanism is thought to be significant to the movement and distribution of many species of fish on the continental shelves where tidal currents are strong.

Humans have developed some strange beliefs about the influences of tides on their lives and behaviour patterns. In Chapter 1 we mentioned the legend that no animal can die unless the tide is ebbing. On the Pacific island of Fiji, the term *matiruku* means both 'low tide in the morning' and someone who is periodically insane, but the reason for the link to local tidal conditions is not known (see also the box on times of the Christian Easter festival at the end of this chapter).

9.6 Mean sea level: the geological record

Some 20 per cent of the earth's surface lies within 100 m above and below the present mean sea level. Large areas of land would be flooded by relatively small increases in sea level (as discussed in Chapter 8) and, conversely, large areas of sea bed would be exposed by a small sea level fall. Fishing trawlers in the North Sea sometimes bring to the surface artefacts of possible human habitation from areas now tens of metres below sea level; similarly, surveys in India have shown evidence for permanent settlements in areas now submerged. In the Mediterranean, ancient harbours such as the old port of Alexandria, Egypt are now submerged as a result of sea level rise over the last 4000 years; some Scandinavian harbours such as Oulu and Vassa in Finland are now inland because of coastal uplift.

Over geological time very significant relative land–sea level changes have occurred. Charles Darwin identified fossil seashells and petrified pine trees in marine sediments in the Andes, 4000 m above present sea levels. The evidence for sea levels having increased relative to the land includes submerged erosion notches, submerged shorelines and deltas. The extension of river valley systems well out across the continental shelf, sometimes as far as a sub-marine canyon at a shelf break, such as

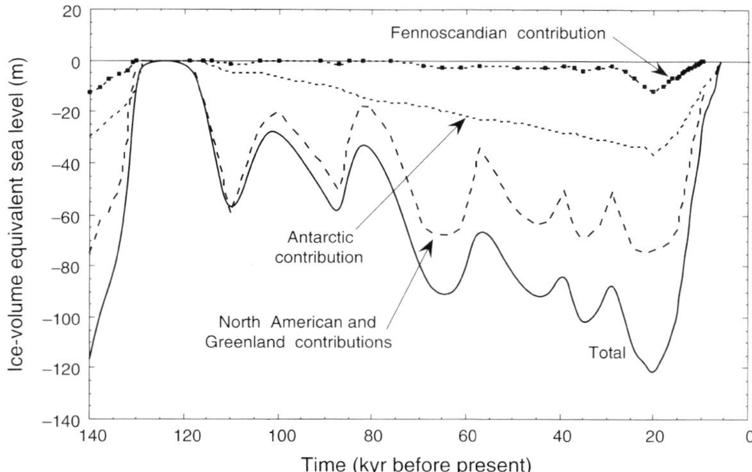

Figure 9.9. Estimates of global sea level change over the last 140 000 years (continuous line) and contributions to this change from the major ice sheets (from IPCC (2001), with permission; supplied by Kurt Lambeck).

the Hudson Canyon off New York, are particularly striking. Submerged forests, peat beds and shells that contain organic material may be dated by radiocarbon methods to determine when they were at mean sea level. Evidence that earlier sea levels were higher relative to the land than today includes raised beaches, elevated coral reefs, tidal flats and salt marshes, as well as wave-cut terraces and sea cave systems in cliffs.

Over very long periods (on the extreme right of Figure 1.3) of over 50–100 million years or longer, sea level changes are due to changes in ocean basin shapes as a result of plate tectonic movements. Over the last two million years, glacial and inter-glacial conditions have alternated on time scales of the order of 10 000 to 100 000 years. During inter-glacials sea levels and climate were similar to those of today. During major periods of glaciation, including the most recent, which was at its maximum 20 000 years ago, large polar ice sheets removed sufficient water from the oceans to lower sea levels by between 100 and 150 m.

Figure 9.9 is an attempt to construct a curve of global sea level changes over the past 140 000 years. About 120 000 years ago sea levels were similar to those of today. The figure shows the reduction in ice volume and corresponding equivalent increase in sea level since then. The curve is based on sea level indicators from a number of localities well removed from former ice sheets, with corrections for isostatic effects. It shows a general pattern of relatively rapid sea level rise from a low level of about −120 m, 20 000 years ago, which gradually slowed down 8000 years ago when levels were some 15 m lower than those today. At times the rise of sea level exceeded 30 mm yr^{-1}. The rise then proceeded more gradually until present levels were reached some 4000 years ago.

Since then, mean sea level changes have been restricted to oscillations of relatively small amplitude.

Modern coasts and estuaries worldwide have evolved during this period of relatively slow sea level change. It seems that in the geological past, changes of MSL of 1 m or more per century have occurred, which is much more than the present rates given in Chapter 7 and more than the projected future rates discussed in Chapter 8.

However, the curve of sea level rise shown in Figure 9.9 is not universal for all coasts, even after correcting for post-glacial adjustment. In many cases there has been a gradual rise over the past few thousand years, but along the Australian margins MSL has fallen by a few metres during the same period. Separate identification of sea level changes and vertical land movements will be possible when we have sufficiently long records of GPS fixing of tide gauge benchmarks in absolute geocentric coordinates (see also Section 7.4).

9.7 Tides past

In the box (Are tides changing?) at the end of Chapter 4 we showed, based on direct sea level measurements, that tidal amplitudes and phases have changed slightly over recent decades and centuries. Tides have also changed over much longer geological periods. There is strong scientific interest in palaeo tides and understanding how energy losses due to tides might have changed and influenced the evolution of the earth–moon system (Section 4.6). If the present rate of increase in the earth–moon separation of 37 mm yr^{-1} were extrapolated backwards, the separation would have been 3700 km less than 100 million years ago. From Equation (2.2) this reduced separation means an increase in the tidal forces and amplitudes of about 3 per cent.

If these increased tidal amplitudes are taken to imply corresponding increases in the energy losses, then further extrapolation suggests that the moon was within a few earth radii of the earth between 1000 and 2000 million years ago. The close proximity indicated by these gross extrapolations, called the Gerstenkorn Event, would imply enormous tides and energy losses. The geological record gives no indication of such an extreme event at any time in the past.

One source of information on ancient tides is found in sedimentary rocks. Careful examination often gives indications of the tidal regime under which the sediments were established. These indications include sand waves, longitudinal sandbanks, interleaved sand and mud sheets, and scoured surfaces. These features can be seen in recent sediments and have also been tentatively identified by geologists in sedimentary rocks from as early as 1500 million years ago. In some circumstances

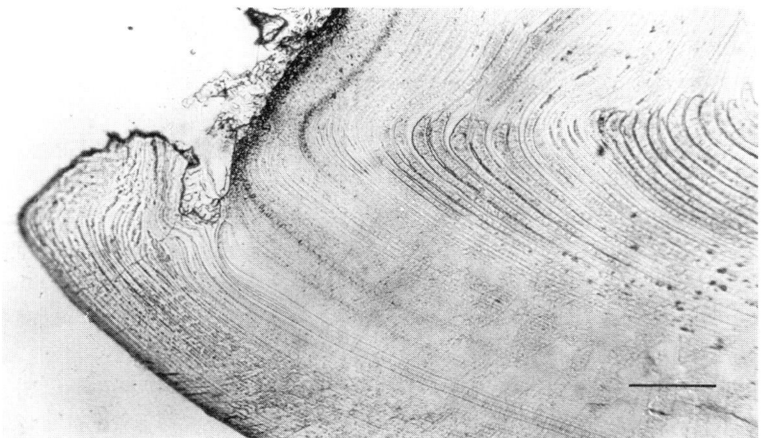

Figure 9.10. Section of the shell of the common European Cockle collected from a position of mean sea level in the Menai Straits, North Wales. Spring and neap cycles have been identified in the banding. The scale bar on the bottom right is 100 μm (supplied by John Simpson).

sand and mud layers are interleaved, indicating periodic changes in the different conditions of sediment transport and deposition; spring–neap modulations are sometimes clearly visible.

An alternative palaeo tidal clock is found in the growth rate of animals, particularly corals, bivalves and stromatolites. These generally show that the number of days in a month and in a year was higher hundreds of millions of years ago when the earth was rotating more rapidly. Spring–neap cycles have been observed in the modern growth of cockleshells *Clinocardium nuttalli* on the Oregon coast of the USA, and the common cockle *Cerastoderma edule* in European waters (Figure 9.10). Similar periodic variations are found in the growth of fossil bivalves. Tides are only one factor in controlling the growth rate and so the results from these indicators must be interpreted with caution.

The absence in the geological record of an extreme Gerstenkorn Event means that assumptions made about past tidal ranges and energy dissipation need further examination. It may be that the near-resonance semidiurnal responses in the present ocean are not typical of previous ocean and continent configurations. Computer models of previous configurations of continental blocks and mid-ocean ridges have been used to investigate this possibility. It appears that during epochs when the continents are in consolidated masses the semidiurnal tides are reduced relative to the longer wavelength diurnal tides. Fragmentation of continents tends to favour the semidiurnal oscillations.

From the Late Carboniferous through to the end of the Jurassic period diurnal tides may have been dominant. Thus over a period from roughly 350 to 100 billion years ago losses due to tidal friction opposing the slower diurnal currents may have been significantly less than those in today's predominately semidiurnal regime. In Section 4.6 and Section 5.6

we showed that most of the tidal energy is dissipated in a few continental shelf areas where there is tidal resonance and large tides. It will be very difficult to identify these areas by attempting to reproduce conditions over different geological epochs.

Tidal conditions in older continental shelf seas, particularly related to MSL increases since the last glacial maximum (Figure 9.9), have been speculatively constructed using estimated water depth and basin geometry together with ideas of amphidrome development and energy losses. A major uncertainty is the way in which the sediments and the tidal basins adjust to different tidal ranges and currents. One interesting area of interaction between geological processes and tidal dynamics is the relationship between basin development and tidal resonance.

The adjustment of continental shelf basins towards or away from resonance as a result of erosion or deposition is a little-studied aspect of tidal dynamics: certainly at times of much lower sea levels, such as those discussed in the previous section, the shelf tides would have been very different because of the smaller basin dimensions of the shallower water depths. Large tidal ranges corresponding to near-resonant conditions would tend to enhance coastal erosion and increase the dimensions of coastal basins. The increase in the natural period of the basin would then lead to a less-resonant response and smaller tidal ranges.

The tides of the Bay of Fundy are a good example of rapid change over only a few thousand years. At present the Fundy tides are near to resonance and so small changes of basin shape could have a large influence. In Chapter 4 we reported apparent increases in $\mathbf{M_2}$ amplitudes at St John Harbor. From computer modelling (Figure 9.11) and from geological evidence it appears that a few thousand years ago the tidal range was only about 40 per cent of the range observed today. From the geological evidence of salt-peat marsh and macrofossils, tree stumps, oysters and boring clams, the onset of amplification was abrupt and rapid some 3400 years ago. The tidal range then expanded at a rate of 0.64 m per century. Computer modelling shows that changes in the depth of water due to MSL increase in the Bay of Fundy and resonant changes are important; changes in the water depth over Georges Bank at the entrance to the Bay controlling the flow of water and tidal energy also have an effect. These changes confirm that on a geological time scale tidal resonances and high rates of energy dissipation are both local and ephemeral phenomena.

9.8 Legal definitions of tidal boundaries

In Chapter 1 we looked at different tidal levels and their use as zero or datum levels. By extension, tidal datums are also used to define

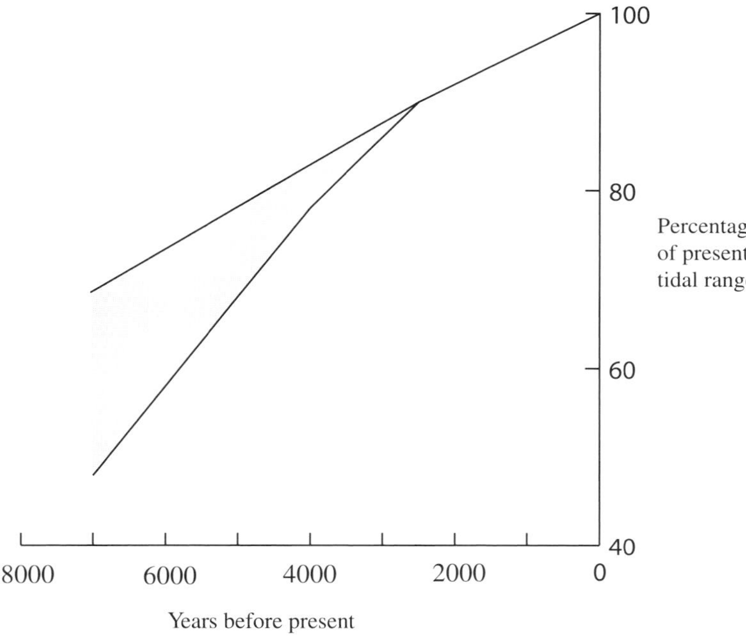

Figure 9.11. Tidal range as a percentage of the present tidal range, averaged over the Gulf of Maine, for the past 7000 years. These results are based on computer modelling with the shaded area showing the range of uncertainty (re-drawn from Scott and Greenberg (1983), with permission).

shorelines which in turn can be adopted as the state, national and international boundaries shown on maps. The determination of these shorelines requires long records of sea level measurements, often a complete nodal cycle of 18.6 years. The clear interpretation of legal definitions calls for an understanding of the tidal characteristics of the region being mapped.

For international boundaries, the United Nations Convention on the Law of the Sea (UNCLOS), which came into force in 1994, requires countries to define a Territorial Sea Baseline. Territorial Seas are measured outwards from this line, usually for 12 nautical miles. UNCLOS (Article 5) defines the baseline as 'the low-water line along the coast as marked on large scale charts officially recognised by the coastal State'. Exactly what is meant by 'low-water' is not more clearly defined, and many countries have elected to use lowest astronomical tide, the datum recommended by the International Hydrographic Organisation for use on navigation charts. Under certain circumstances UNCLOS allows the baseline to jump across bays, river and between islands, as well as along heavily indented areas of coastline.

At the other tidal extreme, UNCLOS defines islands as land features that are 'above water at high tide', but there is no definition, nor a mention of a high tide datum as used on nautical charts; countries that adopt lower high tide datums will obviously define more features as islands, and hence optimise extension of their claim over sea areas.

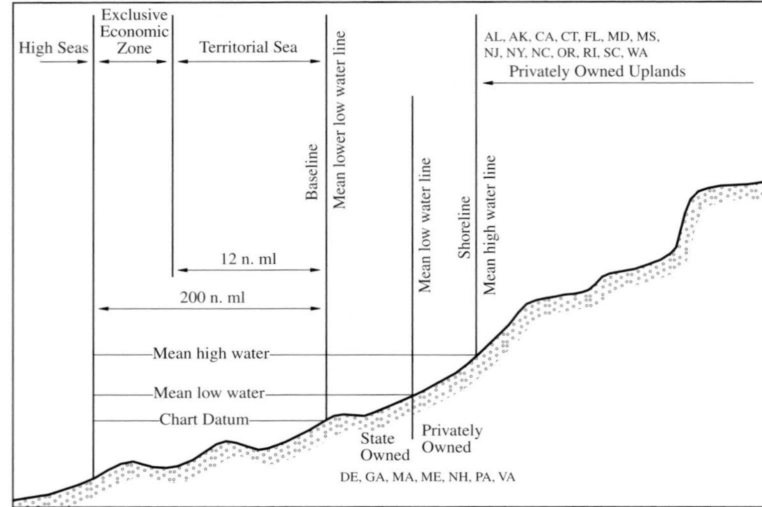

Figure 9.12. Some tidally defined boundaries for different states in the United States. The drawing is not to scale. If the privately owned boundary is near the low water line then beaches are seldom available for public access (based on information provided by NOS/NOAA).

The area of coast covered and uncovered as the tide rises and falls is variously called the inter-tidal zone, the littoral zone, the tidelands or the foreshore. Legal rights and ownership within this zone have been interpreted and defined in different ways at different times in different countries.

In England the Crown has ownership of the foreshore, which extends landwards as far as it is covered by 'the ordinary flux of the sea'. The common-law term 'ordinary high-water mark' has been generally used to describe the boundary between the Crown's foreshore and the adjoining privately owned lands. The national mapping agency, the Ordnance Survey, uses the high and low water marks of a mean or average tide in England and Wales, and of an average spring tide in Scotland.

The law relating to ownership of the tidelands in the United States varies between states (Figure 9.12). Each state is free to adopt its own rules of real property, and questions of ownership are determined under state constitution, statutes and case law. Sixteen coastal states deem the mean high water line to be the private/state boundary. However, seven Atlantic Coast states, Delaware, Georgia, Massachusetts, Maine, New Hampshire, Pennsylvania and Virginia use the mean low water line as the boundary, so beaches can be in private ownership. In Louisiana the boundary is the line of the highest winter tide. Hawaii adheres to its aboriginal, customary concept that the upper reaches of the wash of the waves mark the private/public boundary.

The National Ocean Service charts are the official charts of the United States; they automatically become legal documents in regard to the portrayal of maritime boundaries and tidal datums. Datums are commonly calculated from observations for each National Tidal Datum Epoch; the Epoch is a 19-year period (the nodal cycle), which is reviewed annually for possible revision, and must be actively considered for revision every 25 years. The current Epoch, or 1983–2001, was introduced in 2003. This gradual adjustment allows boundaries to change progressively, for example if there are climate-related MSL or tidal changes.

Why Easter is always a good time to see a tidal bore by moonlight

Easter time varies from year to year, but early or late, Easter weekend is always a good time to see a tidal bore. Spring tides occur when the earth, moon and sun are in line, which is at new and full moon. Very high spring tides follow when the moon and sun are overhead at the equator, near to the 21 March and 23 September equinoxes.

The high spring tides of Easter weekend are inevitable because Easter Day is defined in terms of the lunar cycle and the date of the spring equinox. Easter Day is fixed as the first Sunday after the full moon which happens on, or next after the 21 March spring equinox. It may fall on any one of the 35 days from 22 March to 25 April.

For several centuries after the crucifixion there were arguments about the correct day to celebrate Easter. At the Synod of Whitby in 663/664 the Celtic Church finally accepted the Roman practice, which is used today. In 1963 the second Vatican Council stated that there was no objection in principle to holding Easter on a fixed Sunday. Although a fixed date may serve practical purposes, romantic souls would mourn their annual chance to view an Easter tidal bore under a full moon.

Further reading

All the topics in this chapter have an extensive and growing literature to which the following references can provide an introduction. Dean and Dalrymple (2002) review the engineering aspects of coastal processes; Bird (1993) gives a general account of the development of coastlines worldwide. The relationship between tidal asymmetry and sediment transport is reviewed in Dronkers (1986) and in Friedrichs and Aubrey (1988). Brown and McLachlan (1990), Brown *et al.* (1994), Little and Kitching (1996), Raffaelli and Hawkins (1996) and Weinstein and Kreeger (2001) all discuss various aspects of the ecology of the shoreline. Spawning behaviour of reef fishes is discussed in Robertson *et al.* (1990) and Sale (1991). Further information about the Manhousat people of British Columbia is given in a fascinating account by Ellis and Swan (1981). The sedimentary record of sea level change is described in Coe *et al.* (2003). Recent geological changes of mean sea level are discussed in Lambeck (1980). For a review of palaeo tides see Kagan and Sündermann (1996). Denny and Paine (1998) give an especially interesting account of nodal tidal influences on inter-tidal ecology. Tidal changes in the Bay of Fundy are discussed in Scott and Greenberg (1983). Shalowitz (1962, 1964) remains the authority on coastal boundaries in the United States. A more recent review for international boundaries is given in Antunes (2000). NOAA's National Ocean Service provides a series of publications dealing with the technicalities of defining coastal boundaries, particularly *Tidal Datums and Their Applications* (NOS CO-OPS 1), which is periodically revised.

Questions

9.1 From Equation (9.1) estimate the volume of water exchanged between the ocean and San Francisco Bay in a tidal cycle, given a minimum entrance channel cross-section of 10 000 m^2.

9.2 The currents in a channel are given by (mean flow + $\mathbf{M_2}$ + $\mathbf{M_4}$):

$$U = U_0 + U_2 \cos(2\omega_1 t) + U_4 \cos(4\omega_1 t)$$

The bed-load transport of sediment is proportional to the cube of the current speed, above a critical threshold U_C: Sediment transport = $K(U - U_C)^3$

Construct a spreadsheet:

(a) to calculate the net sediment transport over an $\mathbf{M_2}$ tidal cycle, assuming $U_C = 0.2\ \mathrm{m\,s^{-1}}$ and $U_0 = 0.0$; $U_2 = 1.0\ \mathrm{m\,s^{-1}}$; $U_4 = 0.1\ \mathrm{m\,s^{-1}}$;

(b) as (a) but with a residual current $U_0 = 0.2\ \mathrm{m\,s^{-1}}$;

(c) to investigate the theoretical relationship:

$$\text{Sediment transport } = KU_2^2\left(\tfrac{3}{2}U_0 + \tfrac{3}{4}U_4\right)$$

assuming a zero threshold: $U_C = 0\ \mathrm{m\,s^{-1}}$. The mathematically ambitious should prove this analytically, assuming $U_2 \gg U_0, U_4$.

9.3 A coral reef coastal area dominated by semidiurnal tides has an $\mathbf{S_2}$ phase lag of 180° in local time (g_{S_2}). What does this mean for coral exposure?

9.4 What are the advantages for a country in using lowest astronomical tide for defining the baseline from which to measure Territorial Sea limits?

9.5 If the earth–moon separation is increasing by 37 mm per year, use Table 2.1 to determine how long ago the separation was one earth radius less than at present. Use Equation (2.2) to calculate the percentage increase in average tidal forces at that time.

Appendix 1
Tidal potential

As explained in Chapter 2, the essential elements of a physical understanding of tidal dynamics are contained in Newton's Laws of Motion and in the Principle of Conservation of Mass. For tidal analysis the basics are Newton's Laws of Motion and the Law of Gravitational Attraction. In this appendix we outline the formal mathematical development of gravitational forces and the Equilibrium Tide, based on potential theory. It is an elegant approach, but non-mathematical readers do not need this detail to understand the basic principles of tides and tidal species given in Chapter 2.

The Law of Gravitational Attraction states that, for two particles of masses m_1 and m_2 separated by a distance r, there is a mutual force of attraction:

$$F = G\frac{m_1 m_2}{r^2}$$

We make use of the concept of the *gravitational potential* of a body. Gravitational potential is the work that must be done against the force of gravitational attraction to remove a particle of unit mass to an infinite distance from the body. From potential theory, the potential at P on the earth's surface in Figure A1.1 due to the moon is given by the simple expression:

$$\Omega_P = -\frac{Gm}{MP}$$

Our definition of gravitational potential, involving a negative sign, is the one normally adopted in physics, but there is an alternative convention often used in geodesy, which treats the potential in the above equation as positive. The advantage of the geodetic convention is that an increase in potential on the surface of the earth will result in an increase of the level of the free water surface. Potential has units of $L^2\ T^{-2}$. The advantage of working with gravitational potential is that it is a scalar property, which allows simpler mathematical manipulation. In particular, the gravitational force on a particle of unit mass is given by the gradient of the potential field, $-\text{grad}\,(\Omega_P)$. As a simple analogy, the potential energy of a ball on a mountain depends on its height up the mountain, but the accelerating downhill force on the ball depends on the local slope on the ground. Applying the cosine law to ΔOPM in Figure A1.1

$$MP^2 = r^2 + a^2 - 2ar\ \cos\phi$$

$$\therefore MP = r\left\{1 - 2\frac{a}{r}\cos\phi + \frac{a^2}{r^2}\right\}^{\frac{1}{2}}$$

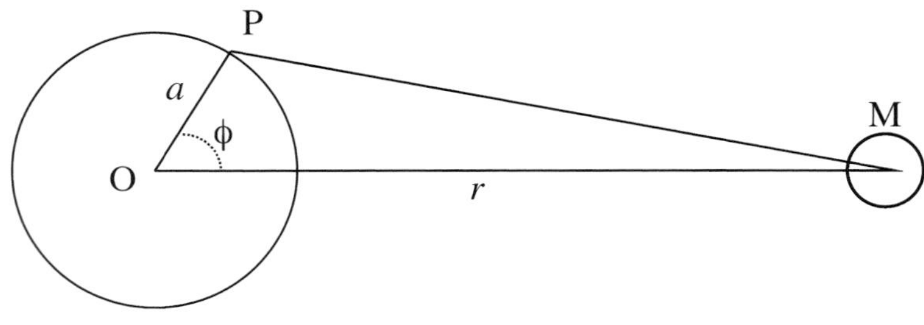

Figure A1.1. The coordinates of the earth–moon system used in this mathematical development.

Hence we have:

$$\Omega_P = -\frac{Gm}{r}\left\{1 - 2\frac{a}{r}\cos\phi + \frac{a^2}{r^2}\right\}^{\frac{1}{2}}$$

This may be expanded as a series of Legendre Polynomials in increasing powers of (a/r):

$$\Omega_P = -\frac{Gm}{r}\left\{1 + \frac{a}{r}P_1(\cos\phi) + \frac{a^2}{r^2}P_2(\cos\phi) + \frac{a^3}{r^3}P_3(\cos\phi) + \cdots\right\}$$

The terms in $P_n(\cos\phi)$ are the Legendre Polynomials:

$$P_1 = \cos\phi$$
$$P_2 = \tfrac{1}{2}(3\cos^2\phi - 1)$$
$$P_3 = \tfrac{1}{2}(5\cos^3\phi - 3\cos\phi)$$

The tidal forces represented by the terms in this potential are calculated from their spatial gradients $-\mathrm{grad}\,(P_n)$. The first term in the equation is constant (except for variations in r) and so produces no force. The second term produces a uniform force parallel to OM because differentiating with respect to $(a\cos\phi)$ yields a gradient of potential which provides the force necessary to produce the acceleration in the earth's orbit towards the centre of mass of the moon–earth system. The third term is the major tide-producing term. For most purposes, because (a/r) is only about $(1/60)$, the fourth term may be neglected, as may all higher terms.

The effective tide-generating potential is therefore written as:

$$\Omega_P = -\frac{1}{2}Gm\frac{a^2}{r^3}(3\cos^2\phi - 1) \tag{A1.1}$$

The force on the unit mass at P may be resolved into two components as functions of ϕ:

- vertically upwards:

$$-\frac{\partial\Omega_P}{\partial a} = 2g\Delta_l\left(\cos^2\phi - \frac{1}{3}\right)$$

- horizontally in the direction of increasing ϕ:

$$-\frac{\partial\Omega_P}{a\,\partial\phi} = -g\Delta_l\sin 2\phi$$

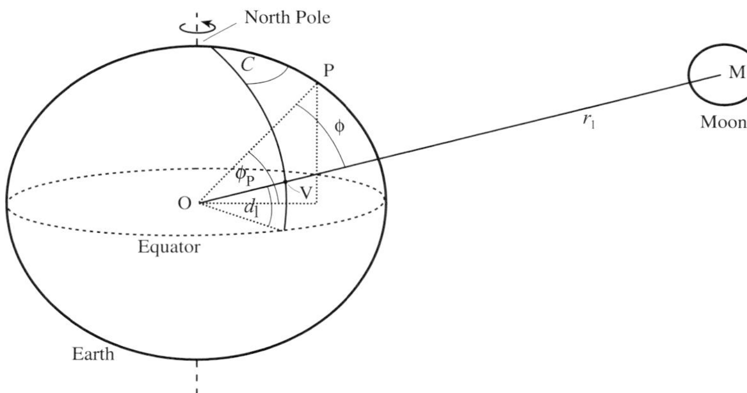

Figure A1.2. The three-dimensional location of a point P on the earth's surface relative to the sub-lunar position. This is a difficult diagram to convert into three dimensions, but it helps to recall that V is on the surface of the earth and on the line O to M.

Δ_l is a constant involving the masses and distances for the system. For the moon:

$$\Delta_l = \frac{3}{2}\frac{m_l}{m_e}\left(\frac{a}{r_l}\right)^3$$

Where m_l is the lunar mass and m_e is the earth mass. r_l, the lunar distance, replaces r. The resulting forces are shown in Figure 2.4.

To generalise in three dimensions, the lunar angle ϕ must be expressed in suitable astronomical variables. These are chosen to be declination of the moon north or south of the equator, d_l; the north or south latitude of P, ϕ_P and the hour angle of the moon C, which is the difference in longitude between the meridian of P and the meridian of the sub-lunar point V on the earth's surface. C moves through a complete cycle in 12 hours and 50 minutes, as the earth rotates. The extra 50 minutes is because the moon has moved on slightly in that time, in its orbit of 27.55 days (see Figure A1.2).

An *Equilibrium Tide* can be computed from Equation (A1.1) by replacing cos ϕ by the expression for the changes in ϕ in the real situation. This expression, which is derived from spherical trigonometry, gives:

$$\cos\phi = \sin\phi_P \sin d_l + \cos\phi_P \cos d_l + \cos C$$

The Equilibrium Tide is defined as the elevation of the sea surface that would be in equilibrium with the tidal forces if the earth were covered with water to such a depth that the response is instantaneous. As explained in Chapter 2 and Chapter 3, it serves as an important reference system for tidal analysis.

The Equilibrium Tide has three coefficients that characterise the three main species of tides:

- the long-period species
- the diurnal species at a frequency of one cycle per day ($\cos C$)
- and the semidiurnal species at two cycles per day ($\cos 2C$).

The Equilibrium Tide due to the sun is expressed in a form analogous to the lunar tide, but with solar mass, distance and declination substituted for lunar parameters (see Section 2.3.2).

Appendix 2
Answers to selected questions

Chapter 1

1.1 Because the period between lunar phases, for example between one full moon and the next, 29.53 days, is slightly less than the length of a calendar month.

1.2 These are the circumference of the earth and the time of its evolution.

1.3 The chart depths are relative to a datum close to Lowest Astronomical Tide (LAT): this is not a horizontal surface. Water moving between two identical depths may in fact be moving vertically in the gravity field.

1.4 (a) needs only about 0.05 m accuracy, and a temporary gauge, but must have datum control;
(b) must be a permanent gauge with real-time data transmission, but 0.05 m accuracy will do;
(c) needs a long-term fixed installation with permanent local benchmarks and 0.01 m accuracy;
(d) is similar to (c), but studies of global change can use satellite altimetry.

Chapter 2

2.1 Around the summer and winter solstices (21 June and 21 December) when the sun's declination is a maximum of 23.5°; the combined diurnal solar and lunar tides when the moon's declination is largest and on the same side of the equator as the sun will give greatest diurnal tidal ranges.

2.2 Newlyn.

2.3 The maximum declinations in March 2002 are 24° 30'S on the 8th and 24° 39'N on the 23rd. Over 18.6 years this varies between 18° 18' and 28° 36'. In 2002 the lunar declination is increasing towards a maximum value, which will be reached in 2006.

2.4 (a) The tidal forces would be reduced to one eighth of the present value; (b) the tidal forces would double; (c) tidal forces would increase by 3 per cent. The assumption is not valid because for a balance of orbital forces, other parameters would change to compensate.

2.5 No work is done because the geoid is a horizontal surface. Although the ship has moved upwards 150 m relative to the ellipsoid of revolution to which the geoid is referred, that ellipsoid is more than 20 km nearer the centre of the earth at the poles than at the equator. The ship will be several kilometres nearer the centre of the earth at the end of the voyage.

2.6 The anomalistic month moves through $360/27.5546 = 13.065^\circ$ per mean solar day. The nodical month moves through $360/27.2122 = 13.229^\circ$ per mean solar day. So they will be in phase every $(360)/(13.229 - 13.065) = 2195$ mean solar days = 6.01 years.

Chapter 3

3.1 Because although they are often dominant, tides are only one component of sea level variability.

3.2 As for sea level measuring systems, the answer depends on the application. Separating the gravitational and radiational solar tides for research studies may need a full response analysis. For routine predictions a harmonic analysis is usually used. For local purposes a simple time difference based on a Reference Port could be sufficient.

3.3 According to the Rayleigh criterion a lunar month will allow independent determination of separate harmonic groups; a full year will allow resolution of individual constituents within each group. The exact number of hours around a month, or a year, is not critical for this resolution.

3.4 At times of maximum lunar declination more tidal energy goes into the diurnal tides and the semidiurnal tides are correspondingly reduced. As the oceans respond more fully to semidiurnal forcing than to diurnal forcing, the total tidal variation of sea level is reduced at these times.

3.5 They tell us nothing about the either the total tidal ranges or the times of high and low water.

3.6 The bottom trace of the contribution of the remaining 58 constituents shows a maximum error of around 0.8 m. Rapid tidal changes at higher harmonics ($\mathbf{M_4}$) are seen around spring tides.

3.7 For $\mathbf{O_1}$

$$\omega = \omega_1 - \omega_2 = 14.492 - 0.549 = 13.943^\circ \text{ per hour}$$

For $\mathbf{S_2}$

$$\omega = 2\omega_1 + 2\omega_2 - 2\omega_3 = 2\omega_0 = 15.0^\circ \text{ per hour}$$

For $\mathbf{K_2}$

$$\omega = 2\omega_1 + 2\omega_2 = 2(14.492 + 0.549) = 30.082^\circ \text{ per hour}$$

3.8 Although due to meteorological changes, because we restrict our analyses to yearly samples, long-term changes of air pressure would appear as changes in $Z_0(t)$. So too would the 18.6-year changes of mean sea level. However, 18.6-year modulations in the amplitude of $\mathbf{M_2}$ would appear as changes in the variance of the tides $\mathbf{T}(\boldsymbol{t})$.

3.9 From the San Francisco predictions in Table 3.4, high water is 5.5 feet at 1.20 p.m.; low water is 1.0 feet at 7.06 p.m.

Chapter 4

4.1 11.79 hours. This value is close to the resonant period for semidiurnal forcing.

4.2 For the spring Equilibrium Tide the amplitude is $1.46H_{\mathrm{M_2}}$; at neaps it is $0.54H_{\mathrm{M_2}}$. Cubing the ratio 1.46/0.54 gives 19.76. In the North Atlantic Ocean the spring and neap tidal amplitudes are $1.33H_{M_2}$ and $0.67H_{M_2}$, which gives a cubed ratio of 7.82.

4.3 Ocean tides have remained essentially constant. However, locally the earlier low levels may have been restricted by harbour drainage. As mean sea levels have increased, so too have high water levels; the low waters can still fall to previous levels, which gives an overall increase in local tidal range.

4.4 The range does not fall to less than 0.6 m at the entrance and the tidal phases are not completely reversed inside and outside the Sound. This more constant response of Long Island Sound to external forcing is because internal tides are restricted by the volume of water that can be exchanged through the entrance during a tidal cycle. At Throgs Neck there is evidence of tidal propagation into the Sound from New York Bay through the East River (smaller range; earlier high water).

4.5 In the ocean the speed of a long wave is 198 m s^{-1}; on the shelf it is 31 m s^{-1}, a difference of 167 ms^{-1}. Diurnal tidal waves have the same speed as semidiurnal waves. Assuming 12.5 and 25 hours for the semidiurnal and diurnal periods, the wavelengths are 1395 and 2790 km respectively on the shelf.

4.6 There is an amphidrome in the shallow sea between the north of Ireland and Scotland.

4.7 Speed = 2.18 m s^{-1}; wavelength = 98 km.

Chapter 5

5.1 0.04 m.

5.2 30.54° per hour and 60° per hour.

5.3 $H_{D_4} = 0.30$ m; $H_{D_6} = 0.25$ m. For a double high water H_{D_4} must be greater than $\frac{1}{4}H_{M_2}$, which it is not. However, $\mathbf{D_6}$ can produce a double high water, as it is greater than $\frac{1}{9}H_{M_2}$.

5.4 At noon, sea level is 1.0 m higher at The Battery than at Kings Point. The corresponding current from The Battery towards Kings Point is 2.2 m s^{-1}. Observed currents will vary because of varying channel width and depths, and the effects of other joining channels. Also, Equation (5.6) assumes steady flow: in practice there will be inertia effects as currents try to adjust to the changing water level gradients.

5.5 The energy available at springs is four times greater than the energy at neaps.

Chapter 6

6.1 ±98 cm at Buenos Aires; ±30 cm at Newlyn. 12 cm at Mombasa is three standard deviations, which is exceeded 1 per cent of the time, but this includes both positive and negative events. For positive events the exceedance is 0.5 per cent.

6.2 114 km h^{-1} is 32 m s^{-1}, the speed of a long wave in water 100 m deep (Equation (4.1)). The result is a hurricane resonance, where the surge generated by reduced air pressure can travel as a wave, at the same speed as the atmospheric system by which it is being generated. Note that this resonance is unusual because it depends only on the water depth and not on any horizontal dimensions of the sea.

6.3 The 12-hour air pressure cycle would produce a harmonic constituent at $\mathbf{S_2}$ due to the inverted barometer effect with amplitude of 1 cm. Maximum sea levels, to coincide with minimum air pressures, would be at 0400 and 1600 local time, giving a phase of 120°. This is called a radiational tide.

6.4 From Equation (6.3) the estimated increase in level is 1.30 m.

6.5 The speed of a long wave is given by Equation (4.1). This applies for both a surge and tidal Kelvin waves. Figure 4.10 shows a phase lag of $(150 - 30) = 120^\circ$ which means that the $\mathbf{M_2}$ tide arrives in Immingham almost four hours after Aberdeen, as will the external surge. In practice this timing will be affected by the interaction between the tide and the surge as they travel together in shallow water.

Chapter 7

7.1 (a) +0.065 m; (b) +0.003 m.

7.2 MTL is greater than MSL by 0.09 m.

7.3 From Table 4.4, $f = 7.29 \times 10^{-5}$ at 30°N. The slope $= 10^{-5}$ and $g = 9.81$ m s^{-2}. Entering these values in Equation 7.1 gives $v = 1.35$ m s^{-1}.
This is an average value, but the Gulf Stream can be concentrated into narrow jets of faster currents

7.4 To quantify ice volume changes over the past 100 years would have required regular surveys from some of the remotest places on earth, which were only beginning to be explored at the beginning of the century. Systematic mapping of ocean temperatures and densities is a very slow process using traditional profiling from research vessels. The extensive observing stage of the World Ocean Circulation Experiment (WOCE) (1990–98) was the first global attempt at mapping the temperatures and salinities at all ocean depths.

7.5 Satellite altimetry can map ice volumes, but the polar ice caps are not reached in the orbits of either TOPEX/Poseidon (66° latitude maximum) or its planned successor missions. Automated surveys by remote underwater vehicles can provide more systematic observations of ocean internal temperature structure, as can the array of ARGO floats now deployed. The WOCE dataset will be a valuable reference against which to detect change.

Chapter 8

8.1 Equation (8.1) can be re-arranged:

$$Q(z) = 1 - (1 - \text{risk})^{1/T_L} = 1 - 0.9^{\frac{1}{100}} = 0.00105$$

which gives a return period of 950 years.

8.2 1/36; 25/36; 10/36.

8.3 1/100; 81/100; 18/100.

8.4 From Figure 8.4, nine 0.1 m bands above 2.0 m have a total percentage occurrence of about 4.5%; we could include half of the band centred at 2.0 m (1.95 to 2.05), giving a total approximate probability of just over 5 per cent.

8.5 Approximately a 10-year return period.

8.6 From Equation (4.1) the wave speed increases from 22.23 ms^{-1} to 22.25 ms^{-1}, or 0.53 per cent. The semidiurnal wavelength is 990 km (Table 4.1), so a quarter-wavelength is 248 km. This increases by 0.53 per cent which is 1.3 km.

Chapter 9

9.1 150×10^6 m^3.

9.2 (a) 0.05K; (b) 0.23K; (c) The complex mathematics required is available on the website of this book.

9.3 The minimum $\mathbf{S_2}$ level occurs at noon, so on spring tides when $\mathbf{M_2}$ and $\mathbf{S_2}$ are in phase, the tides will expose the lowest levels when the sun's heat is at maximum intensity.

9.4 LAT is already generally used on nautical charts as a datum; as the lowest possible low water tidal datum it gives maximum Territorial Sea extension.

9.5 172 million years ago. The decrease in separation from 60.3 to 59.3 earth radii is 1.6%. Tidal forces vary inversely as the cube of the separation, making an increase in tidal forces of 5.1 per cent.

Glossary

AGE OF THE TIDE: old term for the lag between the time of new or full moon (SYZYGY) and the time of maximum spring tidal range. The age in hours is almost the same as the difference in the phases ($g_{S_2} - g_{M_2}$), where the phase difference is in degrees.

AMPHIDROME: a point in the sea where there is zero tidal amplitude due to cancelling of tidal waves. CO-TIDAL LINES radiate from an amphidromic point and CO-RANGE LINES encircle it.

APOGEAN TIDE: tide of reduced range when the moon is near APOGEE.

APOGEE: The point furthest from the earth in the moon's elliptical orbit.

BENCHMARK: *see* TIDE GAUGE BENCHMARK.

BORE: a tidal wave which propagates as a solitary wave with a steep leading edge up certain rivers. Formation is favoured in wedge-shaped shoaling estuaries at times of spring tides. Other local names include eagre (River Trent, England), pororoca (Amazon, Brazil), mascaret (Seine, France).

CHART DATUM: the datum to which levels on a nautical chart and tidal predictions are referred; usually defined in terms of a low water tidal level, which means that the chart datum is not a horizontal surface, but it may be considered so over a limited local area.

CO-RANGE LINES: lines on a tidal chart joining places which have the same tidal range or amplitude; also called co-amplitude lines. Usually drawn for a particular tidal constituent or tidal condition (for example, mean spring tides).

CO-TIDAL LINES: lines on a tidal chart joining places where the tide has the same phase; for example, where high waters occur at the same time. Usually drawn for a particular tidal constituent or tidal condition.

CRITICAL TIDAL LEVEL: a level on the shore where the emersion/submersion tidal characteristics change sharply. Some biologists have suggested that zonation of plants and animals is controlled by a series of such levels, but detailed analysis of tidal statistics shows that the tidal transitions are seldom as sharply defined as the biological boundaries.

CURRENT PROFILE: the detailed variation of current speed and direction between the sea bed and the sea surface.

DATA REDUCTION: the process of checking, calibration and preparation necessary to convert raw measurements into a form suitable for analysis and application.

DECLINATION: the angular distance of an astronomical body north or south of the celestial equator, taken as positive when north and negative when south of the equator. The sun moves through a declinational cycle once a year and the moon moves through a cycle in 27.21 mean solar days. The solar declination varies between 23.5°N and 23.5°S. The cycles of the lunar declination vary in amplitude over an 18.6-year period from 28.5° to 18.5°.

DEGENERATE AMPHIDROME: a terrestrial point on a tidal chart towards which CO-TIDAL LINES appear to converge – 'an imaginary point where nothing happens'.

DIURNAL INEQUALITY: the difference between the heights of the two high waters and of the two low waters of a lunar day.

DIURNAL TIDE: once-daily tidal variations in sea level, which increase with lunar or solar DECLINATION north and south of the equator. When added to semidiurnal tides they can cause a DIURNAL INEQUALITY.

DOUBLE TIDE: a double-headed tide with a high water consisting of two maxima of similar height separated by a small depression (double high water), or a low water consisting of two minima separated by a small elevation (double low water). Also known as an agger or gulder. Examples occur at Southampton, England; along the coast of the Netherlands; and off Cape Cod, in the United States.

EARTH TIDE: tidal movements of the solid earth due to direct gravitational forcing and loading of the adjacent sea bed by marine tides.

EBB CURRENT: the movement of a tidal current away from the shore or down a tidal river or estuary.

ENTRAINMENT: stimulation of an organism by local environmental factors, to adopt a new phase in its cycle of behaviour. For example, a short cold treatment in a laboratory can initiate new tidal rhythms in crabs that have become inactive.

EQUATION OF TIME: the difference between real civil time and apparent solar time, due to the earth's elliptical orbit around the sun.

EQUILIBRIUM TIDE: the hypothetical tide that would be produced by the lunar and solar tidal forces in the absence of ocean constraints and dynamics.

EQUINOXES: the two points in the celestial sphere where the celestial equator intersects the ecliptic; also the times when the sun crosses the equator at these points. The vernal equinox is the point where the sun crosses the equator from north to south and it occurs around 21 March. Celestial longitude is reckoned eastwards from the vernal equinox, which is also known as the 'first point of Aries'. The autumnal equinox occurs around 23 September. The larger solar contribution to the semidiurnal tides leads to large tides at these times, known as equinoctial tides.

EULERIAN CURRENT: flow past a fixed point as measured, for example, by a moored current meter.

EUSTATIC SEA LEVEL CHANGES: worldwide changes of sea level due to the increase in the volume of water in the ocean basins. Volume changes are due to mass increases from melting of grounded ice, and thermal expansion or contraction of the oceans as they warm or cool. Over geological time the shape and volume of the ocean basins themselves also evolve.

FIRST POINT OF ARIES: *see* EQUINOXES.

FLOOD CURRENT: the movement of a tidal current towards the shore or up a tidal river or estuary.

GEOID: the equipotential surface that would be assumed by the sea surface in the absence of tides, water density variations, currents and atmospheric effects. It varies above and below the geometrical ellipsoid of revolution by as much as 100 m due to the uneven distribution of mass within the earth. The mean sea level surface varies about the geoid by typically decimetres, but in some cases by more than a metre.

GLOSS: A worldwide network of sea level gauges, defined and developed under the auspices of the Intergovernmental Oceanographic Commission. Its purpose is to monitor long-term variations in the Global Level Of the Sea Surface, by reporting observations to the Permanent Service for Mean Sea Level (PSMSL).

GPS: a satellite-based Global Positioning System, capable of accurately locating points in a three-dimensional geometric framework.

GRAVITATION, NEWTON'S LAW OF: states that all particles in the universe are attracted to other particles with a force that is proportional to the product of their masses and inversely proportional to their separation apart.

GREENHOUSE EFFECT: the effect, analogous to that which operates in a greenhouse, whereby because of atmospheric effects the earth's surface is maintained at a much higher temperature than that appropriate to balance the incident solar radiation. Solar radiation penetrates the atmosphere, but some of the longer wavelength return radiation is absorbed by the carbon dioxide, ozone, water vapour, trace gases and aerosols in the atmosphere. Observed increases in the concentrations of atmospheric carbon dioxide due to the burning of fossil fuels and other components could lead to a steady increase of global temperatures: the resulting thermal expansion of the oceans and the melting of ice caps would increase sea levels. Concern about possible coastal flooding due to this increase has stimulated recent research into climate dynamics.

GREENWICH MEAN TIME: time expressed with respect to the Greenwich Meridian (zero degrees longitude), often used as a standard for comparisons of global geophysical phenomena, including tidal constituents. More correctly called Universal Standard Time (UST)

GROUPS: *see* HARMONIC ANALYSIS.

HARBOUR OSCILLATIONS: *see* SEICHE and TSUNAMIS.

HARMONIC ANALYSIS: the representation of tidal variations as the sum of several harmonics, each of different period, amplitude and phase. The periods fall into three TIDAL SPECIES, long-period, diurnal and semidiurnal. Each tidal species contains GROUPS of harmonics, which can be separated by analysis of a month of observations. In turn, each group contains CONSTITUENTS, which can be separated by analysis of a year of observations. In shallow water, harmonics are also generated in the third-diurnal, fourth-diurnal and higher species. These constituents can be used for HARMONIC PREDICTION of tides.

HARMONIC PREDICTION: *see* HARMONIC ANALYSIS

HEAD: the difference in water level at either end of a strait, channel, inlet, etc.

HIGH WATER: the maximum water level reached in a tidal cycle.

HIGHER HIGH WATER (HHW): the highest of the high waters (or single high water) of any specified tidal day due to the enhanced diurnal declinational effects of the moon and sun.

HIGHER LOW WATER (HLW): the highest of the low waters of any specified tidal day due to the reducing declinational effects of the moon and sun.

HIGHEST ASTRONOMICAL TIDE (HAT): the highest tidal level that can be predicted to occur under any combination of astronomical conditions.

HYDRAULIC CURRENT: a current in a channel caused by a difference (sometimes called HEAD) in the surface elevation at the two ends. Such a current may be expected in a strait connecting two bodies of water in which the tides differ in time or range. The current in the East River, New York, connecting Long Island Sound and New York Harbour, is an example.

HYDRODYNAMIC LEVELLING: the transfer of survey datum levels by comparing mean sea level at two sites and adjusting them to allow for gradients on the sea surface due to currents, water density, winds and atmospheric pressures.

INDIAN SPRING LOW WATER: a tidal low water datum, designed for regions of mixed tides, which is depressed below mean sea level by the sum of the amplitudes of the principal semidiurnal lunar and solar tides and the principal diurnal tides ($\mathbf{M_2} + \mathbf{S_2} + \mathbf{K_1} + \mathbf{O_1}$); originally developed for parts of the Indian Ocean.

INTERNAL TIDES: tidal waves which propagate at density differences within the ocean. They travel slowly compared with surface gravity waves and have wavelengths of only a few tens of kilometres, but they can have amplitudes of tens of metres. The associated internal currents are termed baroclinic motions.

INVERTED BAROMETER EFFECT: the adjustment of sea level to changes in barometric pressure; in the case of full adjustment, an increase in barometric pressure of 1 mbar corresponds to a fall in sea level of 0.01 m. If there is this full adjustment, the observed pressures at the sea bed are unchanged.

KELVIN WAVE: a long wave in the oceans whose characteristics are altered by the rotation of the earth. Tidal progression is mainly by Kelvin waves. In the northern hemisphere the amplitude of the wave decreases from right to left along the crest, viewed in the direction of wave travel.

LAGRANGIAN MOVEMENT OR CURRENT: the movement through space of a particle of water as measured by drogues or drifting buoys.

LOCAL ESTABLISHMENT or LUNITIDAL INTERVAL: an old term for the interval of time, at a particular location, between the transit (upper or lower) of the moon and the next semidiurnal high water. This varies slightly during a spring–neap tidal cycle. The interval at the times of full and new moon is called 'high water full and change'. *See* NON-HARMONIC TIDAL ANALYSIS.

LONG WAVE: a wave whose wavelength from crest to crest is long compared with the water depth. Tides propagate as long waves. The speed of travel is given by: (water depth $\times$ gravitational acceleration)$^{1/2}$.

LOW WATER: the minimum water level reached in a tidal cycle.

LOWER HIGH WATER (LHW): the lowest of the high waters of any specified tidal day due to the reducing declinational effects of the moon and the sun.

LOWER LOW WATER (LLW): the lowest of the low waters (or single low water) of any specified tidal day due to the enhancing declinational effects of the moon and the sun.

LOWEST ASTRONOMICAL TIDE (LAT): the lowest level which can be predicted to occur under average meteorological conditions and under any combination of astronomical conditions; often used to define CHART DATUM.

LUNAR DAY: the time of the rotation of the earth with respect to the moon. The mean lunar day is approximately 1.035 times the mean solar day.

LUNAR TRANSIT: passage of the moon over the local meridian; when it passes through the observer's meridian it is called the *upper transit*, and when it passed through the same meridian but 180° from the observer's position it is called the *lower transit*.

MAELSTROM: a tidal whirlpool found between the islands of Moskenesy and Mosken in the Lofoten Islands of northern Norway. The term is generally applied to other tidal whirlpools.

MEAN HIGH WATER SPRINGS: average spring tide high water level, averaged over a sufficiently long period.

MEAN HIGHER HIGH WATER: a tidal datum; the average of the higher high water height of each tidal day, averaged over the United States' NATIONAL TIDAL DATUM EPOCH.

MEAN LOW WATER SPRINGS: a tidal datum; the average spring low water level averaged over a sufficiently long period.

MEAN LOWER LOW WATER: a tidal datum; the average of the lower low water height of each tidal day observed over the United States' NATIONAL TIDAL DATUM EPOCH.

MEAN SEA LEVEL: the arithmetic mean of hourly heights observed over some specified period (sometimes 19 years); often used as a datum for geodetic surveys.

MEAN SEA LEVEL TRENDS: changes of mean sea level at a site over long periods of time, typically decades; also called SECULAR changes. Global changes due to the increased volume of ocean water are called EUSTATIC changes; vertical land movements of regional extent are called EPERIOGENIC changes.

MEAN TIDE LEVEL: the arithmetic average of mean high water and mean low water. This level is not identical to mean sea level because of higher harmonics in the tidal constituents.

METEOROLOGICAL RESIDUAL: that part of the observed sea level due to weather effects. *See also* NON-TIDAL RESIDUAL.

METEOROLOGICAL TIDE: *see* RADIATIONAL TIDES.

MIXED TIDE: a tidal regime where both the diurnal and semidiurnal components are significant.

NATIONAL TIDAL DATUM EPOCH: the specific 19-year period adopted by the National Ocean Service as the official time segment over which sea level observations are taken and reduced to obtain mean values for datum definition. The present Epoch is 1983 through 2001. It is reviewed annually for revision and must be actively considered for revision every 25 years.

NEAP TIDES: tides of small range, which occur twice a month at QUADRATURE OF MOON.

NODAL FACTORS: small adjustments to the amplitudes and phases of harmonic tidal constituents to allow for modulations over 18.6 years, the period of a nodal tidal cycle.

NON-HARMONIC TIDAL ANALYSIS: analysis and prediction using traditional methods such as the LOCAL ESTABLISHMENT.

NON-TIDAL RESIDUAL: the part of the observed sea level that is not due to tidal forces. Sometimes assumed to be the same as SURGE or METEOROLOGICAL RESIDUAL, but can include other effects such as seiches and tsunamis.

NUMERICAL MODELLING: calculations of the hydrodynamic responses of real seas and oceans to physical forcing by representing them as a set of discrete connected elements, and solving the hydrodynamic equations for each element in sequence.

PERIGEAN TIDE: tide of increased range when the moon is near PERIGEE. The effect is particularly evident in the tides along the east coast of Canada and the USA.

PERIGEE: the point nearest the earth in the moon's elliptical orbit.

PERMANENT SERVICE FOR MEAN SEA LEVEL (PSMSL): the organisation responsible for collection, analysis, interpretation and publication of mean sea level data from a global network of gauges. PSMSL is one of the Federation of Astronomical and Geophysical Data Analysis Services, under the auspices of the International Council for Science.

POLE TIDE: small variations in sea level due to the Chandler Wobble of the axis of rotation of the earth. This has a period close to 436 days. Maximum amplitudes of more than 30 mm are found in the Gulf of Bothnia, but elsewhere amplitudes are only a few millimetres.

PRESSURE TIDE GAUGES: instruments used to measure the pressure below the sea surface; this pressure may be converted to sea levels if the air pressure, gravitational acceleration and the water density are known.

PROGRESSIVE WAVE: a wave whose travel can be followed by monitoring the movement of the crest. Energy is transmitted but the water particles perform oscillatory motions. *See also* KELVIN WAVE.

QUADRATURE OF MOON: position of the moon when its longitude differs by 90° from the longitude of the sun. The corresponding phases are known as first quarter and last quarter.

RADIATIONAL TIDES: tides generated by regular periodic meteorological forcing. These are principally at annual, daily and twice-daily periods.

RANGE: the difference between high and low water in a tidal cycle. Ranges greater than 4 m are sometimes termed macrotidal and those less than 2 m are termed microtidal. Intermediate ranges are termed mesotidal.

RECTILINEAR CURRENT: *see* REVERSING CURRENT.

RELATIVE SEA LEVEL: sea level measured relative to a local TIDE GAUGE BENCHMARK. Changes include both local vertical land movements, and local sea level changes.

RESONANCE: the phenomenon of the large amplitudes that occur when a physical system is forced at its natural period of oscillation. Tidal resonance occurs when the natural period of an ocean or sea is close to the period of the tidal forcing.

RESPONSE ANALYSIS: the representation of observed tidal variations in terms of the frequency-dependent amplitude and phase responses to input or forcing functions, usually the gravitational potential due to the moon and sun, and the radiational meteorological forcing.

RETURN PERIOD: the average time between events such as the flooding of a particular level. This information may also be expressed as the level that has a particular return period of flooding, for example, 100 years. The inverse of the return period is the statistical probability of an event occurring in any individual year.

REVERSING CURRENT: a tidal current which flows alternately in approximately opposite directions with a SLACK WATER at each reversal of direction. Also known as a RECTILINEAR CURRENT.

REVISED LOCAL REFERENCE: a datum level defined by the PERMANENT SERVICE FOR MEAN SEA LEVEL for each station, relative to which the mean sea level is approximately +7 m. Measurements at the location, over different periods and to different tide gauge benchmarks are all related to this defined datum. The value of 7 m was chosen to avoid confusion with other local datum definitions.

ROTARY CURRENT: a tidal stream that flows continuously with the direction of flow changing through all points of the compass during a tidal cycle. Found away from coastal or shallow-water flow restrictions, where REVERSING CURRENTS are more probable.

SEA LEVEL: the level of the sea after averaging out short-term variations due to wind waves.

SEA LEVEL MEASUREMENTS: may be made in many ways; in all cases some means of averaging out the effects of waves is necessary. For the reading of tide poles the averaging is by eye; in STILLING-WELL GAUGES the waves are damped out by a narrow constriction; acoustic, fixed radar and PRESSURE TIDE GAUGES may apply electronic averaging to rapid samples; satellite altimetry is corrected for general wave conditions within the footprint of the transmission.

SEA LEVEL RISE: long-term increases of mean sea level. The expression is popularly applied to anticipated increases in EUSTATIC SEA LEVEL due to the GREENHOUSE EFFECT and associated global warming.

SEA SURFACE HEIGHT (SSH): the term often used for sea levels measured by satellite altimetry.

SEDIMENT TRANSPORT: the total sediment transported by a current is the sum of the BED LOAD (material partly supported by the bed as it rolls or bounces along) and the SUSPENDED LOAD in the water.

SEICHE: the oscillation of a body of water at its natural period. Coastal measurements of sea level often show seiches with amplitudes of a few centimetres and periods of a few minutes due to oscillations of the local harbour, estuary or bay superimposed on the normal tidal changes.

SEMIDIURNAL TIDES: tidal changes of level twice in a lunar day. Worldwide, semidiurnal tides are the most important because the global ocean is near to resonance at the period of semidiurnal gravitational forcing.

SIDEREAL DAY: the period of rotation of the earth with respect to the vernal equinox (approximately 0.99727 of a mean solar day).

SLACK WATER: the state of a tidal current when its speed is near zero, especially the moment when a reversing (rectilinear) current changes direction and its speed is zero. For a theoretical standing tidal wave, slack water occurs at the times of high and of low water, while for a theoretical progressive tidal wave, slack water occurs mid-way between high and low water.

SPRING TIDES: semidiurnal tides of large range, which occur twice a month, when the moon is new or full.

STANDING WAVE: wave motion in an enclosed or semi-enclosed sea, where the incident and reflected progressive waves combine to give a node of zero tidal amplitude. Maximum tidal amplitudes are found at the head of the basin where reflection occurs. No energy is transmitted in a standing wave, nor is there any progression of the wave pattern.

STERIC LEVEL DIFFERENCES: sea level differences due to differences in water density.

STILLING-WELL GAUGES: instrument system for measuring sea levels; tidal changes of levels are detected by the movement of a float in a well, which is connected to the open sea by a restricted hole or narrow pipe. Wind waves are eliminated by the constriction of the connection.

STOLEN TIDE: a tide that approaches by stealth; particularly, a local term for the high tides that creep up the gullies and marshes of coastal Lincolnshire, England.

STORM TIDE: the sum of a storm surge and an astronomic tide. *See* SURGE OR STORM SURGE.

SURGE or STORM SURGE: a large change in sea level generated by extreme weather conditions, which, if they coincide with high tides, can cause severe coastal flooding. The term is sometimes applied to all meteorological effects on sea level.

SURVEY DATUM: the datum to which levels on land surveys are related; often defined in terms of a MEAN SEA LEVEL. Survey datum is a horizontal surface to within the accuracy of the survey methods.

SYZYGY: astronomical condition of alignment of the earth, moon and sun at new and full moon, the time of maximum spring tidal forcing. Also useful for crossword puzzles.

TIDAL AMPLITUDE: *see* TIDAL RANGE

TIDAL BORE: *see* BORE

TIDAL CONSTITUENT: *see* HARMONIC ANALYSIS.

TIDAL CURRENT: periodic tidal water movement. *See also* TIDAL STREAM.

TIDAL EXCURSION: the LAGRANGIAN MOVEMENT of a water particle during a tidal cycle.

TIDAL PRISM: the volume of water exchanged between a lagoon or estuary and the open sea in the course of a complete tidal cycle.

TIDAL RANGE: the difference between low and high water tidal levels, equal to twice the TIDAL AMPLITUDE.

TIDAL SPECIES: *see* HARMONIC ANALYSIS

TIDAL STREAM: horizontal water movements due to tidal forcing. Also known as TIDAL CURRENT.

TIDE GAUGE BENCHMARK: a stable benchmark near a gauge, to which tide gauge datum is referred. It is connected to local auxiliary benchmarks to check local stability and to guard against accidental damage. The tide gauge datum is a horizontal plane defined at a fixed arbitrary level below a tide gauge benchmark.

TIDES: periodic movements that have a coherent amplitude and phase relationship to some periodic geophysical force.

TSUNAMIS: waves generated by seismic activity, also called seismic sea waves. Tsunamis are also popularly, but inaccurately, called tidal waves. When they reach shallow coastal regions, amplitudes may increase to several metres. The Pacific Ocean is particularly vulnerable to tsunamis.

VANISHING TIDE: periods when the local semidiurnal and diurnal tidal constituents conspire to give several hours of relatively constant sea level.

VERNAL EQUINOX: *see* EQUINOXES.

ZONATION: the pattern of colonisation of the sea shore, whereby individual species flourish in bands associated with particular tidal levels.

References

Aikman, F. and Rao, D. B. (1999). A NOAA perspective on a coastal forecast system. In *Coastal Ocean Prediction*, ed. C. N. K. Mooers. Washington, DC: American Geophysical Union, pp. 467–499.

Antunes, N. S. M. (2000). *The Importance of the Tidal Datum in the Definition of Maritime Limits and Boundaries*. Durham University Maritime Boundaries Research Unit, Vol. 2, No. 7.

Bird, E. C. F. (1993). *Submerging Coasts*. Chichester: John Wiley.

Brown, A. C. and McLachlan, A. (1990). *Ecology of Sandy Shores*. Amsterdam: Elsevier.

Brown, B. E., Dunne, R. P., Scoffin, T. P. and Le Tissier, M. D. A. (1994) Solar damage in intertidal corals. *Mar. Ecol. Prog. Ser.* **105**, 219–230.

Bryant, E. (2001). *Tsunami: The Underrated Hazard*. Cambridge: Cambridge University Press.

Cabanes, C., Cazenave, A. and Le Provost, C. (2001). Sea level rise during past 40 years determined from satellite and in situ observations. *Science* **294**, 840–842.

Cartwright, D. E. (1971). Tides and waves in the vicinity of St Helena. *Phil. Trans. R. Soc. A* **270**, 603–649.

— (1972). Secular changes in the oceanic tides at Brest, 1711–1936. *Geophys. J. R. Astr. Soc.* **30**, 433–449.

— (1999). *Tides, A Scientific History*. Cambridge: Cambridge University Press.

Cartwright, D. E. and Edden, A. C. (1973). Corrected tables of tidal harmonics. *Geophys. J. R. Astr. Soc.* **33**, 253–264.

Chapman, D. C. and Giese, G. S. (2001). Seiches. In *Encyclopaedia of Ocean Sciences*, ed. J. H. Steele, K. K. Turekian and S. A. Thorpe. New York: Academic Press, 6 Vols, pp. 2724–2731.

Church, J. A., Gregory, J. M., Huybrechts, P. *et al.* (2001). Changes in sea level. In *Climate Change 2001: The Scientific Basis*, ed. J. T. Houghton. Cambridge: Cambridge University Press.

Coe, A. L., Bosence, D. W. J., Church, K. D., *et al.* (2003). *The Sedimentary Record of Sea-Level Change*. Cambridge: Cambridge University Press.

Coles, S. (2001). *An Introduction to Statistical Modelling of Extreme Values*. Berlin: Springer-Verlag.

Dawson, W. B. (1920). *The Tides and Tidal Streams, With Illustrative Examples from Canadian Waters*. Ottawa: Department of Naval Services.

Deacon, M. (1997). *Scientists and the Sea, 1650–1900: A Study of Marine Science*, 2nd edn. (Aldershot: Ashgate).

Dean, R. G. and Dalrymple, R. A. (2002). *Coastal Processes With Engineering Applications*. Cambridge: Cambridge University Press.

Defant, A. (1958). *Ebb and Flow*. Ontario: Ambassador Books.

— (1961). *Physical Oceanography*, Vol. II. Oxford: Pergamon Press.

Denny, M. W. and Paine, R. T. (1998). Celestial mechanics, sea-level changes, and intertidal ecology. *Biol. Bull.* **194**, 108–115.

Dixon, M. J. and Tawn, J. A. (1994). *Extreme Sea-levels at the UK A-Class Sites: Site-by-site Analyses*. Proudman Oceanographic Laboratory Internal Document 65.

Dohler, G. (1986). *Tides in Canadian Waters*. Ottawa: Canadian Hydrographic Service, 1986.

Doodson, A. T. (1921). Harmonic development of the tide-generating potential. *Proc. Roy. Soc. A* **100**, 305–329.

Doodson, A. T. and Warburg, H. D. (1941). *Admiralty Manual of Tides*. London: HMSO.

Douglas, B. C., Kearney, M. S. and Leatherman, S. P. (eds.) (2001). *Sea Level Rise: History and Consequences*. San Diego, CA: Associated Press.

Dronkers, J. (1986). Tide-induced residual transport of fine sediment. In *Physics of Shallow Estuaries and Bays*, ed. J. V. D. Kreeke. Berlin: Springer-Verlag, pp. 228–244.

Ellis, D. W. and Swan, L. (1981). *Teachings of the Tides*. Nanaimo, B.C.: Theytus Books.

Flather, R. A. (2000). Existing operational oceanography. *Coast. Eng.* **41**, 13–40.

— (2001). Storm surges. In *Encyclopaedia of Ocean Sciences*, ed. J. H. Steele, K. K. Turekian and S. A. Thorpe. New York: Academic Press, 6 Vols, pp. 2882–2892.

Friedrichs, C. T. and Aubrey, D. G. (1988). Non-linear tidal distortion in shallow well-mixed estuaries: a synthesis. *Est. Coast. Shelf Sci.* **27**, 521–545.

Fu, L.-L. and Cazenave, A. (eds.) (2001). *Satellite Altimetry and Earth Sciences*. New York: Academic Press.

Garrett, C. and Maas, L. R. M. (1993). Tides and their effects. *Oceanus*, **36**:**1**, 27–37.

Giese, G. S. and Chapman, D. C. (1993). Coastal seiches. *Oceanus*, **36**:**1**, 38–46.

Gill, A. E. (1982). *Atmosphere–Ocean Dynamics*. London: Academic Press.

Godin, G. (1972). *The Analysis of Tides*. Liverpool: Liverpool University Press.

— (1992). Possibility of rapid changes in the tide of the Bay of Fundy based on a scrutiny of the records from St John. *Cont. Shelf Res.* **12**, 327–338.

Heaps, N. S. (1967). Storm surges. *Oceanogr. Mar. Biol. Ann. Rev.* **5** (1967), 11–47.

Intergovernmental Oceanographic Commission (IOC) (1985). *Manual on Sea Level Measurements and Interpretation*, Vol. 1, Intergovernmental Oceanographic Commission.

— (1994). *Manual on Sea Level Measurements and Interpretation*, Vol. 2, Intergovernmental Oceanographic Commission.

— (2001). *Manual on Sea Level Measurements and Interpretation*, Vol. 3, Intergovernmental Oceanographic Commission.

Intergovernmental Panel on Climate Change (IPCC) (2001). *Climate Change 2001: Impacts, Adaptation and Vulnerability*. Cambridge: Cambridge University Press.

Jones, I. S. F. and Toba, Y. (eds) (2001). *Wind Stress Over the Ocean*. Cambridge: Cambridge University Press.

Kagan, B. A. and Sündermann, J. (1996). Dissipation of tidal energy, palaeotides, and evolution of the earth–moon system. *Adv. Geophys.* **38**, 179–266.

Knauss, J. A. (1997). *Introduction to Physical Oceanography*, 2nd edn. Englewood Cliffs, NJ: Prentice-Hall.

Lamb, H. (1932). *Hydrodynamics*, 6th edn. Cambridge: Cambridge University Press.

Lambeck, K. (1980). *The Earth's Variable Rotation*. Cambridge: Cambridge University Press.

Le Provost, C., Lyard, F., Molines, J. M., Genco, M. L. and Rabilloud, F. (1998) A hydrodynamic ocean tide model improved by assimilation a satellite altimeter-derived data set. *J. Geophys. Res.* (1998) **103**: **C3**, 5513–5529.

Little, C. and Kitching, J. A. (1996). *The Biology of Rocky Shores*. Oxford: Oxford University Press.

Lubbock, J. W. (1836). On the tides at the port of London. *Phil. Trans. R. Soc.* **126**, 217–266.

Marmer, H. A. (1926). *The Tide*. New York: Appleton.

Mathers, E. L. and Woodworth, P. L. (2001). Departures from the local inverse barometer model observed in altimeter and tide gauge data and in a global barotropic numerical model. *J. Geophys. Res.*, **106**: **C4**, 6957–6972.

Maul, G. A. and Martin, D. M. (1993). Sea level rise at Key West, Florida, 1846–1992: America's longest instrument record? *Geophys. Res. Lett.* **20**, 1955–1958.

Mitrovica, J. X. and Milne, G. A. (2002). On the origin of late Holocene sea-level high stands within equatorial ocean basins. *Quat. Sci. Rev.* **21**, 2179–2190.

Mooers, C. N. K. (ed.) (1999). *Coastal Ocean Prediction*. Washington, DC: American Geophysical Union.

Munk, W. H. and Cartwright, D. E. (1966). Tidal spectroscopy and prediction. *Phil. Trans. R. Soc. A* **259**, 533–581.

Murty, T. S. (1984). *Storm Surges – Meteorological Ocean Tides*. Canadian Bulletin of Fisheries and Aquatic Sciences No. 212. Ottawa: Department of Fisheries and Oceans.

Neumann, G. and Pierson, W. J. (1966). *Principles of Physical Oceanography*. Englewood Cliffs, NJ: Prentice-Hall.

Open University (2000). *Waves, Tides and Shallow-water Processes*, 2nd edn. Oxford: Butterworth-Heinemann/Open University.

Parker, B. B. (ed.) (1991). *Tidal Hydrodynamics*. New York: John Wiley.

Pond, S. and Pickard, G. L. (1995). *Introductory Dynamic Oceanography*, 2nd edn. Oxford: Butterworth-Heinemann.

Proudman, J. (1953). *Dynamical Oceanography*. London: Methuen.

Pugh, D. T. (1981). Tidal amphidrome movement and energy dissipation in the Irish Sea. *Geophys. J. R. Astr. Soc.* **67**, 515–527.

— (1996). *Tides, Surges and Mean Sea-Level*. Chichester: John Wiley.

Pugh, D. T. and Maul, G. A. (1999). Coastal sea level prediction for climate change. In *Coastal Ocean Prediction*, ed. C. N. K. Mooers. Washington, DC: American Geophysical Union.

Pugh, D. T., Hunter, J., Coleman, R. and Watson, C. (2002). A comparison of historical and recent sea level measurements at Port Arthur, Tasmania. *Int. Hydrograph. Rev.* **3: 3**, New Series (November), 27–46.

Raffaelli, D. and Hawkins, S. (1996). *Intertidal Ecology*. Dordrecht: Kluwer.

Redfield, A. C. (1980). *The Tides of the Waters of New England and New York*. Woods Hole, MA: Woods Hole Oceanographic Institution.

Robertson, D. R., Petersen, C. W. and J. D. Brawn. (1990). Lunar reproductive cycles of benthic-brooding reef fishes: reflections of larval biology or adult biology? *Ecol. Monogr.* **60**, 311–329.

Ross, J. C. (1854). On the effect of the pressure of the atmosphere on the mean level of the ocean. *Phil. Trans. R. Soc.* **144**, 285–296.

Roy, A. E. (1988). *Orbital Motion*, 3rd edn. Bristol: Adam Hilger.

Sale, P. F. (1991). The ecology of fishes on coral reefs. New York: Academic Press.

Sandstrom, H. (1980). On the wind-induced sea level changes on the Scotian Shelf. *J. Geophys. Res.* **85**, 461–468.

Schureman, P. (1976). *Manual of Harmonic Analysis and Prediction of Tides*. Washington, DC: United States Government Printing Office.

Scott, D. B. and Greenberg, D. A. (1983). Relative sea-level rise and tidal development in the Fundy tidal system. *Canad. J. Earth Sci.* **20**, 1554–1564.

Shalowitz, A. L. (1962). *Shore and Sea Boundaries*, Vol. 1. Washington, DC: US Department of Commerce, Coast and Geodetic Survey.

— (1964). *Shore and Sea Boundaries*, Vol. 2. Washington, DC: US Department of Commerce, Coast and Geodetic Survey.

Simpson, J. H. (1998). Tidal processes in shelf seas. In *The Sea*, Vol. 10, ed. K. H. Brink and A. R. Robinson. Chichester: John Wiley.

Von Arx, W. S. (1962). *Introduction to Physical Oceanography*. Reading, MA: Addison-Wesley.

Weinstein, M. P. and Kreeger, D. A. (2001). *Concepts and Controversies in Tidal Marsh Ecology*. Dortrecht: Kluwer.

Woodworth, P. L. (1999). High waters at Liverpool since 1768: the UK's longest sea level record. *Geophys. Res. Lett.* **26**, 1589–1592.

Woodworth, P. L., Gregory, J. M. and Nicholls, R. J. (2004). Long term sea level changes and their impacts. In *The Sea,* Vol. 12. Chichester: John Wiley.

World Meteorological Organization (WMO) (2002). *The Science and Forecasting of Tropical Cyclones*. Geneva: WMO Technical Document 1129.

Zanda, L. (1991). The case of Venice. In *Impact of Sea Level Rise on Cities and Regions*, ed. R. Frassetto. Venice: Marsilio Editori.

Index